Is racism an environmental threat?

Debating Race series

David Theo Goldberg, *Are we all postracial yet?*
Ghassan Hage, *Is racism an environmental threat?*
Jonathan Marks, *Is science racist?*

Is racism an environmental threat?

GHASSAN HAGE

polity

The right of Ghassan Hage to be identified as Author of this Work has been asserted in accordance with the UK Copyright, Designs and Patents Act 1988.

First published in 2017 by Polity Press

Reprinted 2017 (twice), 2018, 2109

Polity Press
65 Bridge Street
Cambridge CB2 1UR, UK

Polity Press
350 Main Street
Malden, MA 02148, USA

ISBN-13: 978-0-7456-9226-5
ISBN-13: 978-0-7456-9227-2 (pb)

A catalogue record for this book is available from the British Library.

Library of Congress Cataloging-in-Publication Data
Names: Hage, Ghassan, author.
Title: Is racism an environmental threat? / Ghassan Hage.
Description: Malden, MA : Polity, 2017. | Series: Debating race | Includes bibliographical references and index.
Identifiers: LCCN 2016041398 (print) | LCCN 2017006034 (ebook) | ISBN 9780745692265 (hardback) | ISBN 9780745692272 (paperback) | ISBN 9780745692296 (Mobi) | ISBN 9780745692302 (Epub)
Subjects: LCSH: Racism. | Environmental degradation. | BISAC: SOCIAL SCIENCE / Discrimination & Race Relations.
Classification: LCC HT1521 .H237 2017 (print) | LCC HT1521 (ebook) | DDC 305.8–dc23
LC record available at https://lccn.loc.gov/2016041398

Typeset in 11 on 15 pt Adobe Garamond by Toppan Best-set Premedia Limited
Printed and bound in Great Britain by TJ International Ltd, Padstow

The publisher has used its best endeavours to ensure that the URLs for external websites referred to in this book are correct and active at the time of going to press. However, the publisher has no responsibility for the websites and can make no guarantee that a site will remain live or that the content is or will remain appropriate.

Every effort has been made to trace all copyright holders, but if any have been inadvertently overlooked the publisher will be pleased to include any necessary credits in any subsequent reprint or edition.

For further information on Polity, visit our website: politybooks.com

To Dominique and Aliya

CONTENTS

Preface   viii

Introduction   1

1 Islamophobia and the becoming-wolf of
   the Muslim other   17

2 Islamophobia and the dynamics of ecological
   and colonial overexploitation   52

3 The elementary structures of generalized
   domestication   85

Conclusion: Negotiating the wolf   112

References   134
Index   142

This book, as indicated by the title of the series it belongs to, is about "debating race." It does so, however, in a slightly unusual way. It examines how racism is connected to the ecological crisis. This, I hope, will help us understand both better. Readers of my earlier works on racism will note that some of the book's key concepts, especially that of domestication, have been emerging in my work for many years now (see Hage 1996, 2000, 2003). I have aimed to develop them more fully here. Nonetheless, it should be clear that this book is neither a scholarly work that systematically interacts with existing literature in the fields of racism and environmental studies, nor is it an ethnographic work that can afford to dwell on a wealth of empirical data. There will be enough to direct readers who want either or both of the above. But the text itself is in the essay tradition. I weave together some theoretical and empirical material into what I trust are some coherent propositions about the way racial and ecological questions are related.

The principal proposition is that practices of racial and ecological domination have the same roots. They emanate from what is today the dominant mode of inhabiting and making ourselves viable in the world, what I call "generalized domestication." This concept is central to the book. It is a concept that will, I hope, perform a critical function: it invites readers to start looking at common things slightly differently. It directs them toward an atypical way of experiencing the racial and the ecological domains. And perhaps, on an optimistic note, it might also point a way out of the racial and ecological impasse in which we find ourselves today.

There is a sense in the early twenty-first century that any twentieth-century problem that can be solved has been solved. We are left with spaces of intractable conflicts and impasses. Ireland is behind us and we're left with Israel/Palestine, so to speak. Not many people, then, are game to write about such enduring social problems with a claim to provide "solutions." I certainly am not. The idea that one can have "a solution" to racism, or to the ecological crisis, borders on the ridiculous. But even if we agree with Bruno Latour that "we have never been modern," we might also have to accept that we are incurably "modern enough" in that we cannot help but live in hope. I therefore write expecting that I can leave the reader with a perspective that allows for a richer way of confronting and negotiating those impasses.

There are many people I need to thank in the making of this book, not least the many cohorts of students who have listened to me and engaged with me on the topic of "domestication" since I began deploying it in the early nineties. I am also grateful to the groups of graduates and academics who, over the years, have either endured me because they had to, or generously invited me and offered me a space to develop my argument in Australia, Lebanon, Europe, and the United States. And I cannot but specifically mention all the organizers and participants in "the mother of all racism workshops," the 2014 bus tour of South Africa.

Domestication was the central problematic of a large Australian Research Council grant. This allowed for many engaging discussions with John Cash, Gerhard Hoffstaedter, Michael Jackson, Geoffrey Mead, Gillian Tan, and Henrik Vigh. Special thanks to Geoffrey Mead who has helped with the editing and the referencing. Many other friends/colleagues who have been important either as people who listened to me, or people I have listened to, or both, need also to be thanked: Fadi Bardawil, Lauren Berlant, Niko Besnier, Peter Dwyer, Didier Fassin, Kelly Gillespie, David Theo Goldberg, Karim Makdisi, Saree Makdisi, Achille Mbembe,

Monica Minnegal, Meaghan Morris, Stephen Muecke, Beth Povinelli, Francoise Verges, Eduardo Viveiros de Castro, and Bettina Stoetzer.

Finally, I want to ritualistically (since rituals are important), and non-ritualistically (to emphasize that I am not merely mouthing thanks in a routinized way), thank my family of women starting with my mother May and my sisters Nada and Amale. As always, Caroline Alcorso subjected my imaginative leaps to her healthy empiricism, which does not mean she always managed to curb them. Last but hardly least I want to thank my daughters Dominique and Aliya, who have offered me a gift I have often referred to in my work, and which is crucial to understand this book's conclusion: this is the gift of their mere presence in my life. Since this book is concerned with a future politics, I dedicate it to the two of them.

Introduction

To argue that a social phenomenon is related to the ecological crisis is not difficult today. It is the single most important crisis that has ever faced humanity. When a crisis is deemed "ecological" or "environmental" it is no longer a crisis in a specific relationship to x or y. It becomes a crisis of the very milieu in which we can have relationships to x or y. This was dramatically illustrated by a garbage crisis in Lebanon in 2015. It began as a breakdown in the garbage disposal system due to its complex entanglement with the logic of economic and political sectarian competition in the country. As people began to dispose of their rubbish anywhere they could, the garbage started fouling the already polluted environment. Soon the street smells, the ugly appearance of sea and mountain vistas, the contaminated rivers, permeated everything, causing inconveniences, discomfort and disease. "Garbage disposal" was no longer an unmanageable relation to garbage; it became constitutive of the entire social atmosphere. It affected the way people worked, their mood, where children played, what could be eaten and where one could eat, how and where one could exercise, and more...It is this all-encompassing quality that defines the "environmental crisis" we are facing globally today.

Because of the all-encompassing nature of the crisis exemplified above, not only is it always possible to demonstrate that any social phenomenon is related to the environmental crisis, it is also analytically and even ethically imperative to do so. The crisis makes of the planet a sinking ship. It becomes futile and even obscurantist to study anything aboard the ship on its own, as if the ship is not sinking. Likewise, it becomes equally imperative to show in what way what one is studying can help stop the sinking process. This book is written with these analytical, political, and ethical imperatives in mind. In doing so, I strive to consolidate a political claim that is already taking shape in many activist spaces, and which animal liberationists and eco-feminists began making many years ago, concerning the relation between speciesism and racism: one cannot be anti-racist without being an ecologist today, and vice versa.

But important as it may be to highlight, as I did above, how the ecological crisis affects racism, the book's central argument moves in the opposite causal direction. It aims to explore the way racism itself exacerbates the ecological crisis. This argument takes up a specific form of racism that is quite prevalent today: the anti-Muslim racism generally referred to as Islamophobia. I choose to concentrate on Islamophobia because it is a variant of what remains the most important form of racism pervading the world: the Western/white racism rooted in slavery and colonialism. To be clear, I know that racism is not a disease that only white people can contract. I have witnessed and written about nonwhite racism in a number of places. Nonetheless, if I am to rank various racisms in terms of global impact I have no problem saying that nonwhite racism is far less important than the racism I am examining here. This is true in terms of its empirical frequency, its impact on people, and in terms of its structural effects on and degree of infiltration of existing national and international institutions.

At least since the turn of the century, anti-Muslim practices and beliefs have come to the fore as one of the dominant forms of racism marking our contemporary era. This period saw a globalization of the "Islamic other" around the world. And like all forms of cultural globalization, it involved contradictory processes of homogenization and differentiation (Hannerz 1996). Thus, while an abstract "Islam" and an undifferentiated plural "Muslims" were becoming homogenized as a global threatening form of otherness, the categories that concretely embodied the Islamic threat differed from one country to another. The Muslim other started out by being "Asian" in Britain (there meaning Indians and Pakistanis), "Turkish" in Germany, "North African" in France, "Lebanese" in Australia, and a more vague "Arab" in the United States. But even at these national levels, the picture was rapidly getting more and more complicated during the first decade of the twenty-first century, with more Palestinians and Afghanis, and African Muslims, Arab and non-Arab, joining in almost everywhere. Today, Syrian and Iraqi nationals have also been added to the "Muslim otherly mix" as a result of the flow of refugees escaping the Syrian and Iraqi wars. What's more, the rise of first Al-Qaedah and then ISIS accelerated a global diffusion of transnational Muslim radicalism among immigrant and Western-born subjects of varied ethnic and national backgrounds, adding an "icing of otherness" on the racialized Muslim cake.

The racist practices that accompanied this globalization have been many and are increasing. Statistical and anecdotal data are abundant and can be obtained from many international and national organizations around the world monitoring the incidence of attacks against Muslims or people thought to be Muslim. There are many cases of Muslim women being shouted at and abused, or having their hijabs ripped off, in the streets or on public transport. These practices, which were already prevalent in the late twentieth century, well before 9/11, have become far more numerous throughout the Western world today. There are also increased reports of Muslim job-seekers being denied jobs the moment their Muslim background is suspected. Likewise Muslim workers are joked about, humiliated, and discriminated against at their workplaces. More publicly, Muslims have to contend with the refusal to respect their taboos in cases well-known internationally, such as the Muhammad cartoons or the attacks on halal meat. They are faced with the continual belittling of the loss of lives among them, whether in war zones such as Syria, Iraq, or Afghanistan, or in Israel, or at sea among those seeking asylum. They are also faced with a Western electioneering culture where Muslim-bashing has become de rigueur and widely seen by politicians as a route to popular success. Every time Muslims turn on a television or read the newspaper they have to come to terms with the prejudice peddled by the media in all its diversity. At the same time, they have to live with a routinization of the spectacle of seeing themselves or their fellow Muslims behind barbed wire, either in jail or in refugee camps, waiting on one or another European border, or in detention centers such as those built by the Australian government in and outside of

Australia. These scenes and practices come in addition to the more common and unspectacular everyday petty forms of racialization and marginalization that mark the phenomenon everywhere: interactions laced with avoidance, disapproval, aggression, or hatred. The totality of these images and practices creates an inescapable feeling among many that today "Muslim" is short for "the wretched of the earth."

Despite all this, there are some analysts who want to differentiate Islamophobia from racism "proper." They do so in the name of a tighter and more rigorous definition grounded in the history and logic of racism as it emerges in the early phases of modernity. According to such writers it is not useful to talk about race and racism unless we are dealing with a mode of thinking that espouses some form of biological conception of race. For the purposes of this work, such approaches have all the hallmarks of what Pierre Bourdieu critically identifies as a form of scholastic thought (Bourdieu 2000: 49). "Scholastic" here refers to a mode of thinking that detaches racism from its practical/usage context and conceives it as an academic exercise aimed at some kind of pure knowledge, a desire to classify for classification's sake.

This intellectualist tendency has had a limiting effect on both anti-racist analysis and anti-racist politics.

Indeed, if we are to compare racism and anti-racism across history, we can say that racism has exhibited a far greater malleability than academic anti-racism. It has morphed, and shown a capacity to target a variety of people, sometimes many at the same time: blacks, Asians, Arabs, Jews, Roma, and Muslims. It has been deployed as part of a technology of segregation, of conditional integration, and, most dramatically, of extermination. It has efficiently constructed its object, successfully adapting to the dominant modes of classification of the time, be they phenotypical, biological, cultural, or a combination of these and more. Comparatively speaking, academic anti-racism has become conceptually somewhat ossified and is always trying to catch up with the racists' fluid mode of classification.

Whereas racists happily move from one form of racism to another, caring little about logical contradictions, inconsistencies, and discrepancies in their arguments, anti-racist academics spend an inordinate amount of time trying to judge racists on precisely such grounds. They criticize racists as if the racists are students or fellow academics with whom they are having disagreements in a tutorial room about how to interpret reality. The lethal performativity of racism, which is what is most important to the racists, is given far less attention than needed. It is as if the racists' greatest sin is that they are bad thinkers: they are "essentialists," they deviate from "classical biological racism," or they make false empirical statements about reality that the anti-racist academics work for long hours to correct by highlighting a lot of statistical data that proves them incorrect.

In the case that concerns us, for instance, Islamophobic classifications vaguely and continuously fluctuate between the Arab, the Muslim, and "Islam," between the racial phenotypes, the ethnic stereotypes, and the religious generalizations. That is, from the perspective of the racializing subject, it is unclear where the Arab and the Muslim begin and end, where they are separate and where they fuse and where they even go beyond to delineate anyone who in the eyes of the Western racists looks like a "third-world-looking-person." Both racists and the police, on the lookout for potential "Muslim terrorists," have killed or captured South Americans, Africans, Sikhs, Hindus, Greeks, Southern Italians, and many others. Keeping to the vagueness of racist thought is crucial since it conveys something important about the imaginary nature of the experience itself. Nor is this vagueness, in fact, a problem from a practical perspective. Racists have always managed to be exceptionally efficient specifically by being vague. It could even be said that vagueness, empirical "all-over-the-place-ness," contradiction, blocking-of-the-obvious, and even sometimes a totally surrealist grasp of reality, are the very conditions of possibility of the maximal efficiency of racist practices. All of this is missed if what we concentrate on is that the racist is using an empirically incorrect classification.

This is of particular importance when studying the relation between racism and speciesism, as we will do here. Animals (and sometimes plants too), as they are classified and imagined in the process of domesticating nature, have long served as a source of metaphors for the portrayal of subjugated and inferiorized people. The analysis of these metaphors puts us immediately face to face with the poverty of scholastic "empirical anti-racism." When faced with the metaphoric space where racists think of Blacks as monkeys, or of Jews as snakes, or of Muslims as cockroaches or wolves – classificatory metaphors that are continuously changing and fluctuating depending on time and place – does it mean anything to spend even one second trying to "empirically" prove, "Hang on a second, Jews are not snakes," or "Muslims are not wolves"? Clearly not, and to be sure, no one does it. The argument that needs to be made, however, is that these animalistic metaphors reveal the nature of all racist classifications, whether they are explicitly metaphoric or not. When facing animalistic metaphors, rather than querying their truth claims we are invited to ask more important questions, such as "What does the imaginary of the Jewish other or the Muslim other as dog, snake, hyena, or wolf tell us about the racists themselves, about their sense of power, and about their practical dispositions toward their other?" My point is that this should also be the kind of question we ask when faced with nonmetaphoric and pseudo-empirical statements such as "Shari'a law is going to take over the country," rather than spending time showing that the claim does not make empirical sense.

In fact, racialized animalistic categories can sometimes tell us more about racism as an everyday practical orientation toward the other than the more general intellectual definitions of racism. This is because the metaphors embody a practical orientation. They are carriers of a "manual" with complete "what to do" instructions. Take for instance the classical association of racism with inferiority and inferiorization. While this association is largely obvious and true, and has important consequences, it does not necessarily, by itself, orient the racist person toward a particular action. "Inferior" can mean many things. Sometimes the explicit classification of someone as inferior hides the fact that the classifier fears them and thinks of them as superior, at least in some way. Less ambivalently, people can classify someone or something as "inferior" and still think they are very loveable and cute. Indeed babies can be thought of by adults as "inferior." In a different way, most patriarchal thought also sees women as "inferior" while also portraying them as objects of beauty, love, and care. Without knowing what imaginary is articulated to the notion of inferiority, it is hard to know its exact practical signification. When a slave is referred to as a "bull," however, or a woman servant a "lamb" this gives us a better access to the practical racial imaginary behind such words. We learn from their "bullification" or their "lambification" what is desirable, possible, and preferable to do with them. Likewise, we know what a racist wants to do when they associate a "Jew" with a "snake" or a "virus." The resultant action to be taken is much clearer than when she is declared "inferior." The animalistic metaphor is not just an "observational racist category" but a declaration of intent.

This book is thus more interested with how well racists do what they do rather than how intellectually rigorous they are. As far as my argument is concerned it is good enough to call "racist" any bundle of practices which aim at problematizing, excluding, marginalizing, discriminating against, rendering insecure, exploiting, criminalizing, and terrorizing and harbouring exterminatory fantasies against an identity group of people imagined as sharing a common and inheritable determining feature. This can be considered an extension of what Hannah Arendt calls race-thinking (Arendt 1944; see also Razack 2008), as long as we keep in mind that there is no race-thinking that is not at the same time race-practicing.

Approaching racism as a practical reality also foregrounds the question of perspective. Racism involves at least two practical experiences: the experience of the perpetrator and the experience of those subjected to it. Many analyses of racism do not sufficiently distinguish the two. Like for those philosophers who imagine a "pure knowledge" produced by an "eye that must not have any direction," whom Nietzsche (1997) warned against, the object of analysis is imagined as a perspective-free thing called "racism." In contrast to this, the argument in this work is strongly perspectival. It concentrates on exploring the experience and life-world of the perpetrators. In doing so, however, it is taken for granted that the subject who expresses Islamophobia or acts in an Islamophobic manner is never "either Islamophobic or not." Islamophobia is more often than not a dimension of such a subject's existence: it colors his or her thinking and practices in a variety of forms, degrees, and intensities. An empirical sociology of Islamophobic subjects would aim to examine these differences and the variables that affect them. This is not, however, what I am aiming to do in this book. Here, I am more interested in Islamophobia and Islamophobic culture as a mode of being in its generality. What does it mean to experience the world Islamophobically? What modes of belonging to one's environment and surroundings does Islamophobia presuppose? What kind of subject is summoned by it? What kind of classification of otherness and what kinds of practices does it generate? And perhaps most important, how does one stage the viability of one's life Islamophobically? That is, how do some subjects come to think that in acting Islamophobically they are protecting what makes their lives worth living? The book's key argument is that answering these questions brings us face to face with one of the most important ways in which racism constitutes an environmental threat.

There are many ways in which racism can have an impact on the environmental crisis. A whole subfield, environmental racism, is devoted to questions of racial equity and justice in the production and distribution of the harmful effects of the ecological crisis (Bullard 1993). More often than not, in such literature, racism, like any other social process, is conceived as autonomous from and external to the environmental phenomena that it affects. That is, racism is seen as unfolding in one space and the environmental crisis in another, and the two happen to meet. This follows a general analytic differentiation between human-nature and intra-human relations. Thus, while changes in the practices of hunting, mining, or growing wheat are seen as having a direct ecological dimension, practices as diverse as sexism and taxation are not. Racism is seen as a variant of the latter sort of practice. This book concentrates on arguments that position racism in the former sort. It wants to argue that the racial crisis manifested by Islamophobia and the ecological crisis not only happen to have an effect on each other; they are in effect one and the same crisis, a crisis in the dominant mode of inhabiting the world that both racial and ecological domination reproduces. Thus, racism is an environmental threat not just because it has here and there an impact on the environmental crisis from the outside, which it has, but because it intensifies it from within. This is a paradoxical claim in that it is at the same time a weaker and a stronger causal claim. It is weaker in that it is not about the direct causal effects of racist practices on the environment. But it is stronger in that it highlights a more fundamental causality. Racism is an environmental threat because it reinforces and reproduces the dominance of the basic social structures that are behind the generation of the environmental crisis – which are the structures behind its own generation.

The book is organized in the form of a search for what these foundational structures are and the ways racism works to validate and consolidate them. It begins by noting the many areas of similarity between the ecological crisis and the crisis expressed in Islamophobia. It then proceeds to examine where the roots of these similarities lie. Three interrelated domains are explored. These are located first in the forms of othering, domination, and governmentality shared by racism and environmental domination; second, in the dynamic of capitalist exploitation of people and resources entailed by colonialism; and third, in the structures of "generalized domestication" as a mode of existence. The book consists of a chapter on each of these domains of similarity. It argues that it is neither in the mode of governmentality, domination, and othering explored in chapter one, nor in colonial capitalist exploitation examined in chapter two, but in "generalized domestication" as developed in chapter three that the most fundamental generative structures of both racism and speciesism can be found. It is this generalized domestication that racism reproduces, and in so doing also reproduces and revitalizes the harmful modes of relating to the environment that plague our world today. The concluding chapter thinks through the possibility of alternative modes of relating to human and natural otherness less dominated by generalized domestication.

ONE

Islamophobia and the becoming-wolf of the Muslim other

There is a long tradition of establishing a relation of analogy between human domination and the domination of nature by highlighting similarities between the two practices, and between the modes of thinking and classification associated with them. Indeed, it is impossible for a student of race and racism not to come face to face with what Marjorie Spiegel, in reference to slavery, has called "The Dreaded Comparison" (Spiegel 1996): the similarity between the micro and macro forms of racist subjugation of people and the speciesist subjugation of animals. "Micro" and "macro" here mean similarities at the level of individual practices and at the level of apparatuses of subjugation. This highlighting of resemblances has been made throughout history. It has one of its earliest "analytic" expressions in Aristotle's well-known comparison, which is still worth quoting at length given its all-encompassing and examplary nature:

> It is clear that the rule of the soul over the body, and of the mind and the rational element over the

passionate, is natural and expedient; whereas the equality of the two or the rule of the inferior is always hurtful. The same holds good for animals in relation to men.... Again, the male is by nature superior, and the female inferior; and the one rules, and the other is ruled; this principle of necessity extends to all mankind. Where then there is such a difference as that between soul and body, or between men and animals (as is the case of those whose business is to use their bodies, and who can do nothing better), the lower sort are by nature slaves, and it is better for them as for all inferiors that they should be under the rule of the master. For he who can be, and therefore is, another's and he who participates in rational principle enough to apprehend, but not to have such a principle, is a slave by nature. Whereas the lower animals cannot even apprehend such a principle; they obey their instincts. And indeed the use made of slaves and of tame animals is not very different; for both with their bodies minister to the needs of life. (Aristotle 2014: 1990)

The same analytic comparisons continue to be made today, though without Aristotle's approving attitude, which our current sensibilities find morally distasteful.

From a different angle, Bob Miles, a central figure in the sociology of racism, examined the way racist modes of thinking had already originated in the West in the modes of categorizing the working classes as exemplified in the work of Le Bon (Miles 1993). But Le Bon in his *Les lois psychologiques de l'évolution des peuples* (1894) did not only show the similarity between race and class differences. He also extended it to differences between men and women. Between the two, he argued, was an "infranchissable abîme." Le Bon backs his argument with well-established "racializing" technologies: "The average of the skulls of female Parisians classes among them the smallest skulls with which we are acquainted, almost on a level with the skulls of Chinese women, and scarcely above the feminine skulls of New Caledonia" (Le Bon in Todorov 1994: 114). Not surprisingly his argument also articulates similarities to human-animal differences. He argued, "The lowest strata of the European societies are homologous with the primitive men" and that "it would suffice ... to allow time to act to see the superior grades of a population separated intellectually from the inferior grades by a distance as great as that which separates the white man from the negro, or even the negro from the monkey" (Le Bon in Todorov 1994: 113).

But, as is clear to anyone who has heard racists speak or has read them, the comparisons are not only "analytical." People who aim to dominate, racially or sexually, mobilize animal categories as part of their technology of domination. As David Theo Goldberg points out: "Monkeys, baboons, orangutans, and mules have been central media of racially characterized dehumanization, the projection of degraded intelligence and delimited rational capacity. Animalization and bestialization have long been integral to the history of racist representation, perhaps even more so than the objectification or 'thingi-fication' of people" (Goldberg 2015: 48–9).

Colette Guillaumin makes a strong critical usage of this comparative logic in her bodily conceptualization of sexism and the institution of marriage, which she refers to as *sexage*. As she puts it:

> The reduction to the state of a thing, more or less admitted or known about in relations of slavery or serfdom, exists today in industrialized urban centres, under our very eyes, dissimulated/exposed in marriage, an institutionalized social relationship, if ever there was one. But the idea that a class is used (literally: manipu-lated like a tool), that is, treated like a cow or a reaper, is in the very progressive minds of our contemporaries supposed to be ascribable to past ages or to despotisms as oriental as they are primitive, or at best to be the expression of a provocative cynicism. (Guillaumin 1995: 193)

Vandana Shiva points out (1989, 1991) that it is not only the labor of slaves or women's labor that traditionally gets subsumed "by definition" into nature. So does the labor of all colonized non-Western, nonwhite people.

But it isn't the case that the comparison between the intra-human and human-animal domain is only made in the cases of racial and animal domination that involves mistreatment and violence. Comparisons regarding "gentler practices" are also found throughout history and continue to be heard today. Thus the late sixteenth-century saying mentioned by Keith Thomas: "A just man will not over-toil the poor dumb creature, nor suffer it to want food or looking to. But if he be so pitiful to his beast, much more is he merciful to his servants, his children and his wife" (1983: 155). Likewise there is a clear similarity in the logic of preservation present in the establishment of nature reserves for plants and animals and the logic present in multiculturalism (for an interesting articulation of the two logics see Cárdenas 2012).

In all of the above, what is conceived as a relation between the racial and the ecological is a relation of analogy established by an analyst or by racists themselves. When the analogy is made by the racist person it is meant as a "putdown." Calling another person an "ape" with the intention of abusing them proceeds from the belief that an ape is a lower order of life than humans. The person on the receiving end does not need to share the views of the racist that an ape is a lower order of life to feel hurt. A knowledge of the intention to hurt is enough. But if they do share their views, which is often the case, they are clearly doubly hurt. A number of people subjected to racism exclaim "They treated me like a dog," agreeing that this is a demeaning mode of being treated. Very rarely would we have an interaction where the racialized person questions the pertinence of the metaphor directed at them and tells the racist, "What's wrong with dogs or apes? They are better than humans in many regards."

When the analogy is made analytically and critically, such as it is done by Colette Guillaumin, the highlighting of the animalistic metaphor aims to reveal the vileness of the intra-human relation of domination and exploitation. It is assumed by the analyst that the intra-human relation is not seen for what it is and the comparison with a human-animal relation helps to make its nature clearer. Some of the analogies I developed in writing *White Nation* (Hage 2000) were aimed to work in this way. I will use a passage that was edited out of the book here, as it captures well the critical spirit in which analogies between multicultural forms of racial domination and the domination of animals are made:

> Multiculturalism stands to assimilation in the way free-range chooks [Australian informal term for chickens] stand to battery [i.e., caged] hens. Free-range chooks are certainly, as the classification implies, freer than battery hens and living a healthier and happier life, and one can say that in being permitted to roam about "on the range" they are in fact allowed "to retain their own culture." Battery hens, on the other hand, since they are required to fit into a restrictive life in cages, are submitted to a kind of assimilation to a dominant culture that is not theirs. Nonetheless it should be remembered that neither process of farming chickens has the interest of the animal other as its final aim. Indeed, the aim of both processes is the extraction of nutritional value from the chicken and the production of a specific kind of eggs and meat for a specific kind of market: The first, free-range, for a classier, more cosmopolitan and discerning consumer, and the second, caging, for a mass of supposedly undifferentiating consumers. In much the same way, we can say that neither multiculturalism nor assimilation have the interest of the ethnic/racial other as their final aim. They are both modes of producing/transforming other cultural forms into something acceptable for consumption by the dominant culture (ethnic food, ethnic art, ethnic

labor). The first produces ethnicity for a refined market of cosmopolitan consumers of cultural otherness, the second produces an otherness made bland such as to not disturb the cultural sensitivities of a presumed undifferentiating and unsophisticated majority.

Here, as in Guillaumin's text, the analysis is located in the critical sociological tradition of unearthing or revealing hidden relations of domination. It is presumed that we humans worry less about hiding the relations of power and exploitation, or the instrumental reasoning, that underlie our relationship with animals than we do when it comes to our relations with other humans. We think that there is no moral problem, or at least less of a moral problem, in being instrumental when dealing with "nature." But we don't feel the same when it comes to our dealing with each other. We tend to experience enough guilt to want to hide the instrumental reasoning and calculative rationality that governs intra-human relations, or as in the case of slavery, work hard on convincing ourselves that the intra-human relations before us are not actually intra-human relations. Domination and instrumental reasoning deployed intra-humanly toward each other are experienced in an ambivalent way. On the one hand these are pragmatically seen as necessary, and on the other they are felt to be a symbolic equivalent of cannibalism that we are ashamed of. It is like the mythical tale of the survivors of an airplane crash on a remote mountain top. Some end up having to eat the others but the survivors prefer not to talk about it. They act as if it has not happened. Thus, to show that a mode of intra-human relations that posits itself to be domination-free or instrumentalism-free is similar to an instrumental human-animal relation can help highlight the hidden dimensions of instrumentality and power in such relations.

Useful as it might be, there are however many limits and problems with dwelling for too long in this analogic domain. Animal liberation activists, and academics working on human-animal relations, have rightly argued that this metaphoric usage of human-animal relations to highlight problems in intra-human relations can itself be a form of instrumentalism that belittles the experience of animals by making what happens to them useful only insofar as it helps us understand what happens to humans. Donna Haraway's critique of Derrida's encounter with his cat as he describes it in *The Animal that Therefore I Am* (2008) is exemplary in this regard (Haraway 2008). Looking at an over-instrumentalized ox and saying, "See, this is what it is like to be a Vietnamese migrant working in a sweatshop" might help us understand a little bit better what happens to immigrants in a sweatshop, but it also involves an intellectual instrumentalization of the ox that shows a certain blindness to its experience and suffering. Paradoxically, this intellectual blindness is not unlike the racialized blindness to the suffering of the sweatshop worker that the analogy is trying to highlight in the first place.

Just as important for our purposes, establishing a conceptual analogy and showing similarities and resemblances between two processes of domination does not imply the existence of an actual relation between them. It does nonetheless open the way to show as Val Plumwood's investigation of the "logic of domination" did:

> the way in which different kinds of domination act as models, support and reinforcement, for one another, and the way in which the same conceptual structure of domination reappears in very different inferiorized groups: as we have seen, it marks women, nature, "primitive people," slaves, animals, manual laborers, "savages," people of colour – all supposedly "closer to the animals." (Plumwood 1993: 29)

Consequently, even if they don't indicate actual relations in themselves, let alone the existence of a single generative mechanism behind them, similarities do help us to establish the intellectual grounds for thinking the possibility of such relations and such mechanisms. It is with this in mind that this chapter explores some of the ways in which the racial crisis expressed in Islamophobia resembles the ecological crisis today. To fully appreciate this resemblance we need to explore the colonial nature of Islamophobia.

Islamophobia as colonial racism

To say that Islamophobia is a form of colonial racism is to say two things: First, that Islamophobic practices and conceptions of otherness have been molded by the history of colonialism, and second, that they continue to function as a technique of governing racialized people in order to reproduce the basic racial colonial structure that still underlies a large part of the world even in this neocolonial age. As Claude Lévi-Strauss has explained, despite the unraveling of the classical European empires, we still live in societies

> that are the outcome of a historical process which has made the larger part of mankind subservient to the other, and during which millions of innocent human beings have had their resources plundered and their

27

> institutions and beliefs destroyed, whilst they them-
> selves were ruthlessly killed, thrown into bondage, and
> contaminated by diseases they were unable to resist.
> (Lévi-Strauss 1966: 126)

That this is a racialized division hardly needs to be demonstrated. Nor that the many racisms, whether against blacks or against Asians, it gave rise to were part and parcel of the apparatus that governed this division. Likewise with anti-Muslim racism. Stereotypes of Muslims appeared well before the history of colonialism. They were nonetheless reshaped and retooled to function as part of the colonial governmental apparatus deployed to dominate and exploit colonized Muslim lands and people (Rodinson 1978; Said 1978; Norman 1980; Shryoc 2010; see also Esposito and Kalin 2011; Arjana 2015).

Foregrounding this is important because there is a long and strong liberal tradition that thinks of racism as being itself the crisis. Here "crisis" is imagined as an aberration of an assumed nice, racism-free, democratic, and egalitarian normality. This, however, is as absurd as arriving in the midst of a Southern US plantation and thinking that the racism of the white slave-owners was an indication of a crisis. Quite the opposite. A "thriving" racism is an indication of a lack of crisis. The crisis is when racism fails to do its governmental job. It is when a slave or a colonized person refuses to work, or to accept his or her dehumanization, or refuses to "know his or her place," and so on, that racism is in crisis. We can say that American anti-black racism carries a trace of crisis within it since the abolition of slavery. While still mobilized to govern black bodies, it carries within it a fear of failure that has only increased since the gains of the civil rights movements and even more so with the election of Barack Obama as president. It is a racism always tinged with a sense of panic: it wants to govern racialized bodies and yet is worried that it is failing to do its job properly. It ambivalently marks a sense of entitlement to power over the racialized body and a fear that it is losing that power. As we shall see, when compared to past colonial anti-Muslim racism, it is precisely such a crisis that is reflected by Islamophobia today.

The becoming-wolf of the Muslim other

A racist conspiracy theory circulated in the aftermath of the terrorist attack of September 11. It maintained that Arabs/Muslims were too dumb to plan something like this; it was really Mossad that was behind the attack.

This "theory" combines two racial stereotypes: the anti-Semitic stereotype of the conspiratorial Jew and the stereotype of the dumb colonized other. The fact that the two are played against each other to frame an action by Arab Muslims points to the ambivalent position that the latter have always had within the spectrum of European colonial racist classification of otherness.

This spectrum was, and to a certain extent continues to be, structured by a mind/body polarity. On one pole we have the other of the mind/will who is the devious and scheming other, the manipulative other, the conspiratorial other, the other that can thwart our plans and undermine us, the other who, deep down, we fear might be superior to us . . . at least with regard to cunning and craftiness. On the other pole we have the other of the body: an unambiguously inferior other, inferior in terms of intelligence, inferior in terms of technical know-how, inferior in terms of capacity to be productive (the category of "the lazy other," for instance). To be sure, the racialized other is never totally either an "other of the will" or an "other of the body" but is usually various combinations of both. The combination varies, and continues to vary according to a multiplicity of historical and social variables. Nonetheless, there were some relatively stable "ideal-types." While the epitome of the other of the will was the anti-Semitic figure of the Jew, the quintessential other of the body was the slavery infused colonial construction of the black African. The first was predominantly perceived as an exterminable other while the second was more often than not seen as an exploitable other. This does not mean, of course, that the other of the body was never exterminated. These classifications have always fluctuated. Colonial history is full of examples of colonized others who were initially deemed exploitable but then became perceived as dangerous and exterminable. (Steinmetz's (2007) wonderfully nuanced analysis of the multiplicity and complexity of German colonial classifications is a good example.)

It is within the field constituted by the polarity between the racism of exploitation and the racism of extermination, and their imaginary others, that we can place the often ambiguous categorization of the Arab/ Muslim in early European colonial classification. For the Arab/Muslim is at the same time, a hybrid that is both a "Jew" and a "black African" and an in-between that is neither. He is an other of the will and the body, and neither; too lazy and cunning to work, and yet too dumb to be cunning. Even in imaginary racial phenotypical terms, the Arab/Muslim is neither a "Jewish type" nor a "black type" but an in-between, something fluctuating between neither and "both." The Arab/

Muslim is both an uncivilized other like the black African but also, like the Jew, belonging to an early modern civilization that has shaped European civilization. The Arab/Muslim is a monotheist like the Jewish other but with an "inferior" religious imagination (e.g., paradise as full of virgins) akin to what the colonizers thought of as more primitive religious forms. And while the dangerous Jew of the anti-Semitic imagination is categorized as a snake, a hyena, or vermin, and the exploitable black as a domesticated animal, the Arab is neither.

During the colonial era proper, the Arab/Muslim seemed to be for the European colonizer something polluting and nonexploitable, what the pig is for Jewish and Muslim culture. While not threatening in the way the "Jewish snake" was, the Arab was threatening in the sense of being unreliable and shifty as well as an abject nuisance: more cockroach than snake. It is this kind of ambivalence that produces the ugly stereotypes of racist anti-Arab literature such as what one still finds in Zionist novels exemplified by the work of Daniel Easterman (a pen name of Denis M. MacEoin). Here is his portrayal of an Arab: "He was small, thickset, with shifty eyes and a furtive manner. He was the sort that masturbated without enjoying it. David guessed that he fantasized about fifteen-stone women with massive breasts and pouting lips called Fatima" (Easterman 1987). It is a similar imaginary of the Arab that one finds in the more political depiction of them as cockroaches by Zionist colonists.

While today's Arab of the racist imagination, particularly the Bin Laden-like and now ISIS-like figure of the scheming international terrorist, is a continuation within a global context of the above racializing dichotomy, a new dimension has been added to the equation with the Arab/Muslim being increasingly conceived as a threat. While the metaphor of the cockroach is still alive and well, it must nonetheless contend with the rising image of the Arab/Muslim as "wolf." Such changes in racialized imagery are never absolute and differ from one place to another. Still there is no doubt that there is an increasing identification of the racialized Muslim with the figure of the wolf. Interestingly, it is only during the Algerian war of independence that French conceptions of the Arab other contain notable wolfish portrayals (see for example the famous "loup de l'Akfadou" [Saadi 2010]).

In the Western imaginary, the wolf is the ultimate representative of the threatening undomesticated other of nature. Throughout history, wolves' ungovernability gives them a mythic-imaginary dimension that finds its way into popular culture and slowly but surely eclipses their more objective portrayal by naturalists. Indeed, the objectively minded person who would venture to convince people that their conceptions of the size and the menace of wolves are exaggerated, would have been as brave and Sisyphean as the anti-racist trying to correct the racists' empirical grasp of the threat of the other today. As the sense of their ungovernability grew, wolves were publicly portrayed as much larger than they actually are; they grew to become giants in the seventeenth- and eighteenth-century American and European public imaginary. They are also always imagined as far stronger than they are, endowed with supernatural strength. Those of us who have grown up in, or in the shadows of, European culture have inherited such an imaginary through the various European tales that embody the myth of the wolf: we all know that they can, in no time, gobble our grandmothers, not to mention our most beautiful and innocent girls.

Today, even when wolves are being rehabilitated in various programs of "re-wilding," the wolf continues to provide us with a key metaphor to define those who threaten our sovereign domain. It is this wolf that is being mobilized today to racialize Muslims: Google "lone wolf" and note the innumerable ways it is used to mark the fear of an attack by uncontrollable Muslim individuals sympathetic to ISIS. But as was the case with real wolves, the very idea of a "lone wolf" is contentious. Some in the US are not happy with it:

> How many "lone wolves" must attack the West for being free and open, before people realize: These aren't lone wolves at all. To paraphrase Ian Fleming (or John Adams): One Muslim terrorist is a lone wolf. Two are called a pair of lone wolves. Three or more are called a pack of wolves. That's what we have here: a pack of wolves. (Sodahead 2014)

On exactly the same day and on the other side of the Atlantic somebody was saying exactly the same thing to David Cameron:

> Britain's Prime Minister David Cameron responded to the terrorist attack in Sydney, Australia, in which two of more than 17 hostages were killed by an Islamic extremist. Referring to "lone wolf terrorists," he said this was a reminder of the danger that Britain faces. It would seem to me that we have had enough Lone Wolves to constitute a substantial wolf pack. (Korol 2014)

Indeed one can easily say about Muslims as perceived in the West today what the ethno-zoologist and wolf specialist Geneviève Carbone notes about wolves in Western history. There were many facts and many legends circulating about wolves, but "between facts and legends, fear built an empire of exaggeration where they became the eternal accused" (Carbone 1991: 14). The emotionally loaded and reasonably phantasmagorical images of the ferocious and murderous wolf became an integral part of the European reality of the wolf. Carbone notes that a war strategist advising Louis XIV on how to defeat the English seriously proposed this plan: "a wolf eats a man in two days, unload ten thousand wolves across the Channel and in no time, there will not be a single Englishman left" (p. 16). Maybe a similar imaginary haunts the Brexiting English as they look at all the asylum seekers amassed across the Channel in Calais.

While there are many ways of being threatening that the metaphor of the wolf can stand for, they all lead us to what is perhaps the most crucial component in the colonial imaginary of the Muslim other today: his or her ungovernability. In the first instance, this ungovernability emerges vis-à-vis two sets of laws and policies that are intimately linked with the global government of the neocolonized other today. The first set comprises the international laws that enshrine national sovereignties over national borders. These are supported by immigration policies that regulate the movement of people across these borders. They are the primary means through which a nation-state governs the transnational mobility of nationals and nonnationals into and outside of its space. The second set contains the social and cultural settlement policies that govern the mode through which nonnationals and their offsprings are inserted/assimilated (socially or culturally) into national space. The first set of laws aims at a politics of containment: that is, like the fences governing the spatial allotment of various animals on a farm, borders and immigration laws aim at restricting people's movement to their respective nation-states. The second set of laws aims at governing social integration. The "Muslim" emerges as ungovernable today vis-à-vis both sets of laws. S/he appears as both uncontainable and impossible to integrate.

The uncontainable other

For most Middle Easterners, the Sykes-Picot agreement (1916) exemplified the authoritarianism and the arbitrariness with which colonial borders were established. The borders the agreement created separated pre-existing national worlds, fragmenting and regimenting once-continuous territories and communities. Like the fences of a well-maintained farm, they initially worked to limit people's movements and in time also to limit the imaginary space of their aspirations. They controlled people's fantasies of a viable life as much as they worked to control and regulate population flows between nation-states, particularly between the former colonies and the West. Today, however, they have become more like badly maintained fences less and less able to perform their spatial function, with people increasingly crisscrossing them chaotically. It would be a mistake, however, to think that this is only a question of weak or even obsolete national borders.

When looking at the way people have experienced national borders around the globe, analysts often do not notice that borders have given rise to a global apartheid. Apartheid is above all an institutionalization of two realities structured by race and class within a single space. It is always assumed that such a space is national. But the colonial order of the border – the borders established by Western colonizers – functioned as a global apartheid structure dividing the whole world into two realities, where race and class combined to define two coexisting and yet separate spaces whereby the quality of life, of infrastructure, health, and mobility, differed radically.

Two types of borders divide the global world. The first and most obvious type is the national border, which separates different nation-states. The second and less obvious type is a racialized class border, which separates two different experiences of mobility in the world of national borders. In apartheid fashion, these two experiences delineate two separate realities or worlds that coexist within the same global space. On the one hand, we have a world where a "third-world-looking" transnational working-class and underclass citizens live, and are made to feel that national borders are exceptionally important and difficult to cross. In this world, visas, checkpoints, searches, investigations, interviews, immigration bureaucracies, refusal of permission to cross, language problems, embassy queues, fences, cost of travel, and the like all combine to make national borders appear salient and important realities. On the other hand, we have a world experienced as open, in which people move smoothly across national borders, experiencing the world as almost borderless. This is the experience enjoyed by the largely White upper classes, who are made to feel truly at home in the world. This explains the paradoxical experience, which Wendy Brown (2010) refers to, of increased spatial openness and fusion accompanied by a proliferation of protective walls. Some people roam the globe like masters, others like slaves. Some are the subjects of the global order, others are its objects, often circulating strictly according to the needs of capital. Some are "expatriates" and others are "immigrants." This bifurcated experience extends a long history of two global realities that differentiated the world of the slave owners from that of the slaves and the world of the colonial masters from that of the colonized laborers.

Here we face something important about the current refugees. As colonially traced borders begin to crumble under the weight of internally and externally produced crisis, asylum seekers try to muster a bit of agency in the face of the global and national forces that aim to keep them "nationally sequestered" where they are, despite the chaos, death, and destruction that surrounds them. They are like escaping slaves attempting to free themselves from the increasingly suffocating national order of things by traveling in its shadows and through its cracks. And indeed, when caught, they are increasingly caged and treated like escaped slaves. But in becoming the maroons of the enslaving order of national borders, they are endangering more than the already collapsing national borders; they are also endangering its global apartheid structure. That is, when they are stopped from crossing a border, as when the Australian navy stops a boat of asylum seekers from entering Australian waters, we think that they are only stopped from transgressing Australian national borders. At one level they are. But at the same time, and more importantly, they are also being stopped from transgressing and crossing the class/apartheid border between the world where national borders are made to be so extremely important and where they are enslaved and forced to remain, and the borderless world of smooth global sailing that remains the reserve of the economic and cultural upper classes.

The unintegrable other

If multiculturalism and assimilation are the two broad policies with which the West today governs the social integration of the cultural other within it, the figure of the Muslim stands out for the difficulties with which either of these policies appears to work. In many parts of the world multiculturalism is portrayed as an alternative to monocultural assimilation. This, to a certain extent, is obviously true. But it also obscures the fact that multicultural governance always relies on the continued existence of an assimilationist tendency in its very heart in order to achieve its aims. All western governmental documents on multiculturalism celebrate diversity but ensure with a "but" or an "as long as" that no one forgets that this diversity should not happen at the cost of national cohesion, core values, and so on. Assimilationism, therefore, always exists as a disciplinary technique deployed specifically to ensure that the diverse cultures that are integrated into the multicultural fold are "good to integrate and be multicultural about" in the first place. If multiculturalism is deployed to teach multicultural horses how to be mounted by those who wanted to "enjoy" their culture, assimilation is the technique deployed to "break them" before such a training. It is in this sense that we can realistically speak of a multicultural-assimilationist apparatus. Those who are polemically inclined should note that I am not saying that this is necessarily a bad policy because of this. Perhaps this combination of multiculturalism and assimilation is the most appropriate policy that a government can take. Nonetheless, it remains worthwhile to note analytically that the very condition of governmental efficiency of this combined assimilationist-multicultural apparatus requires the fostering at the same time of an ideological polarity between the two. What has marked the relation of this apparatus to the Muslim other is that in many instances neither side of the polarity worked.

The first element that contributed to a conception of "the Muslim" as outside the multicultural realm is the existence among them of a substantial and increasing number of pious people. To be pious here does not simply mean going frequently to the mosque or holding intense religious beliefs. More important, it means considering all aspects of one's everyday life as governed by the laws of one's God (Brague 2007). It is this kind of religiosity, particularly because it is not the Christian kind, that constitutes a serious negation of the logic of multiculturalism. Multiculturalism can be defined by an ability to find room for minor elements of "the law of the other" to exist within the dominant national law – here I don't necessarily mean "law" in a formal sense, though it could be, but more an anthropological conception of law as "the way of doing things." It constitutes what can be called a relation of encompassment. The dominant national law opens a space where the law of the other can be encompassed by the national law. What is encompassed can vary in content and in magnitude but what cannot possibly change is that the dominant national culture has to be the encompassing culture.

The problem that Western multicultural states have with pious Muslims is that what the latter see as their laws are nothing short of the laws of God. The idea that the dominant national law encompasses a space where you can speak your language, eat your food, and follow your kinship rituals for as long as you understand that this is a space offered to you by the dominant culture is relatively unproblematic. But the idea of having that dominant national law offer a space for the laws of God is sacrilegious in the eyes of the pious. Indeed, for people who take their religion seriously, the situation is reversed. It is the laws of God that are the all-encompassing ones and the laws of the nation that are the minor ones. The very relation of encompassing-encompassed cultures on which multiculturalism is based is here inverted and intimations of ungovernability start to arise. But this is not where they end. That some Muslims think of themselves as belonging to a politicized transnational community or *umma* has given a further earthly flavor to this mode of living under the law of God, transforming it into a mode of trans-national belonging that freely fluctuates between the metaphysical and the political and further undermines the very idea of national governmentality.

What has made the difference are the anti-Western political tendencies that articulated themselves to this transnationalism. This began following the Iranian revolution when Iran instituted for the first time a rule of law that openly portrayed itself as a kind of transcendent Muslim political will. This political will exercised itself for the first time transnationally with the Salman

Rushdie affair. Suddenly Middle Eastern Muslims were acting as if they were in a position to openly "sentence" a person living in and subject to the protection of the law of a Western nation-state. Even more threatening to the Western national law/will, numerous Muslims who were supposed to be its subjects showed themselves to be the willing agents/subjects of this transnational Muslim will by volunteering to carry out the sentence themselves. Since that time, there have been many occasions where Muslims have shown themselves to be the subjects of a transnational will that is other than the West. The rise of al-Qaedah and then ISIS as proponents of transnational Islamic terrorism firmly put a collectively imagined "Muslims" in a position of enmity to the dominant culture and outside the realm of what multicultural governmentality can deal with.

The Arab/Muslim asylum seeker arrives at the doorsteps of the West always already a transgressor of national and class borders and a socially and culturally ungovernable body hovering between states of "cockroachness" and "wolfishness." It is in further analyzing the significance of the state of ungovernability captured by these metaphors that we reach the point where the similarity between the colonial and the ecological crisis is at its most instructive.

The politics of ungovernable waste

In categorizing something as "ungovernable" one can easily slip into thinking that ungovernability is an intrinsic quality of an object. Clearly, however, this is never the case. Ungovernability is a relation. When something is ungovernable it reflects some of its qualities in relation to the capacity of the apparatus trying to govern it. What is ungovernable for one apparatus does not have to be so for another. The same goes for the cockroach-ness and wolfish-ness of a racialized other. This is another reason for taking, as argued earlier, the naturalistic metaphors of racism seriously. These metaphors not only tell us something about the racialized, they give us access to the imaginary of power, or to what Nietzsche (1997: 84) calls the "sense of power" of the racists themselves. Think of the change noted above between the racialization of the Muslim as a "cockroach" and as a "wolf." Not only does the difference between wolf and cockroach tell us something about the threatening nature of the Muslim other, it also tells us something about the racist's own sense of power: one gazes down at cockroaches, not at wolves. For Nietzsche a sense of power is the way one experiences the amount of power one has and how to deploy it. For example, two people can be equally powerful, but the first might be confident that their power is rising while the second feels that their power is declining. These two people will have a different sense of power despite having the same "amount" of power. The person who senses that their power is increasing will yield their power more benevolently perhaps than the person who feels that their power is waning who might be likely to deploy their power more cruelly. Let us now, with this in mind, go back to the category of the Muslim cockroach/wolf and the way it reflects a threat of ungovernability.

As noted earlier, the racialization of Arab/Muslims in the colonial era rarely aimed to make of them an exploited labor force. This tendency was reinforced with the discovery of oil. The colonial economies of oil extraction did not require the kind of labor necessary in the colonial economies of mining (Mitchell 2013). And while s/he was considered to be scheming, the Arab was not seen as scheming in the threatening way the Jews were seen by the fascists. The usage of categories such as "cockroach" reflected both the European experience of them as a practical nuisance and, more existentially, as "abject" and "in between" in a classic Mary Douglas (1966) "matter out of place" way. Indeed, what the racialization of the Arabs always intimated was that they were more of the order of dirt, rubbish and waste, an inevitable left-over of the process of colonization that one has to live with and manage but that one can do without.

In a way, the racialization of the Arabs as "waste" prefigures the current forms of neoliberal racialization where the uselessness of the racialized to the racist displaces the earlier racism where the figure of the useful laborer or slave dominated. Today, racism is dominated by neither the plantation nor the mine. It has all the hallmarks of the processes of waste management in the ways it is institutionalized in the forms of mass incarceration in the United States, in the besiegement of Gaza, or in Australia's detention centers. But if the category of the cockroach can point us to the Arab as a form of waste, the category of the wolf doesn't immediately do so. It only does so in the context of ungovernability. "Wolfish waste" is the waste that starts as an ungovernable object but slowly takes us into an ungovernable/wolfish environment. One can see a remarkable similarity, for example, in the language describing the increasingly threatening agglomerations of plastic waste floating in the oceans and Muslim asylum seekers floating around in those same oceans. Not surprisingly, it is through this imaginary of the Muslim as "ungovernable waste" that Islamophobia points us most clearly to the ecological crisis as it is experienced today. For what is the ecological crisis if not a crisis of ungovernable waste, whether in the form of plastic in the oceans, toxic chemicals in the rivers and the soil, or greenhouse gases in the atmosphere?

For a long period we humans have felt reasonably secure in our capacity to utilize nature and its resources. We felt that resources were either endless or able to regenerate themselves. And while we knew that there were all kind of toxic and polluting waste materials that resulted from both the extraction of resources and their consumption,we felt we were able to manage, recycle, or at least live with this waste. The ecological crisis began to intrude into our lives as crisis precisely at the point when we started experiencing the results of industry's and government's loss of control and inability to manage and recycle waste in the ways we hoped for, giving rise to an ungoverned overflow of unrecyclable waste that is increasingly polluting – visually, chemically, and in many other ways – our lands and waters as well as the atmosphere. Furthermore, this waste has affected and is increasingly affecting some of the basic constitutive dimensions of our life on earth in fundamental ways. We find ourselves unable to contain its effects or rectify the problem. This is the case with the increasingly frequent extreme weather events consequent on the rise in global temperatures. It is in this sense that the ecological crisis takes the form of a practical impasse of governmentality: we are at a loss as to what can be done.

This chapter has shown that a similarly structured ungovernability is also the key feature of the colonial crisis expressed by Islamophobia today. While all colonial racisms embody a certain fear of the colonized subject, none forefront this "phobia" the way Islamophobia does. It is a phobia born out of the inability of anti-Muslim racism to do its colonial job of governing its Muslim object. Unlike those who use this phobia to say that Islamophobia is a form of prejudice distinct from racism (see Perbner 2005), what needs to be stressed is how it makes it a distinct form of racism: Islamophobia equals anti-Muslim racism plus the fear that anti-Muslim racism is not doing its job. It is this that makes it a late colonial artefact: an expression of a colonizing power that remains powerful enough to colonize but that is nonetheless in a state of decline, fearing that it is losing its grip over what it has historically dominated. In so doing it exhibits the same patterns and dynamics of the ecological crisis today, which also reflects a power to dominate nature that is losing its capacity to govern and yet is still aiming to dominate – while fearing what it has historically dominated.

I began this chapter by arguing that my point is not merely to establish the similarity between the racial and ecological but to take that similarity as a starting point, in order to think of the deeper generative mechanism behind both. In a sense, "waste" already points us in this direction. This is because, while the processes of governmentality we have concentrated on in this chapter help us understand how racial and ecological waste goes beyond control, governmentality and domination do not explain how waste is actually produced. To do so we need to examine the processes of extraction, exploitation, and consumption of which waste is a byproduct. This is what puts us in front of what many consider the most important mechanism producing both the ecological crisis and colonial racism: the mechanism of capitalist exploitation and accumulation. In the next chapter we will explore to what extent capitalism constitutes the shared generating structure we are after, the one by which colonial racism, in reproducing, confirms itself as an environmental threat.

TWO

Islamophobia and the dynamics of ecological and colonial overexploitation

In reflecting on the relation between racism and crisis in the previous chapter I highlighted the fact that racism was not an aberration of an otherwise egalitarian and democratic space. It was part of *another* governmentality directed at subjects whose lives are constructed as less valuable in themselves, and against whom more repressive and violent forms of subjugation can be deployed with less difficulty. Agamben's "homo sacer" (Agamben 1998; 2005) and Achille Mbembe's "necro-politics" (Mbembe 2003) have also conveyed this governmental differentiation in dialogue with Foucault's conceptualization of biopolitics (2008). Interestingly for us a similar argument has also been advanced from an inter-speciesist perspective. As Jeffrey Nealon (2016: 3) asked concerning the way the sacrificing of plants and animals forms the public secret of any biopolitics: "under a regime dominated by an intense concern for 'life,' why do we live and they die?"

In many ways, this very question points us to what is important about the differential between biopolitics and its necropolitical negation; it also shows us what is left out. For what is elided in the question "why do we live and they die?" is the question that in a way already contains the answer, which is "why do they die so that we live?" Formulated in this way the question immediately takes us from the domain of differential governability to the domain of extraction and exploitation, which is more present in Mbembe than among those who have deployed the term "necropolitics" after him.

This is crucial to highlight. In Marxist theory the concept of exploitation operates as an explanation and a critique of the concept of inequality. "Inequality" seems to imply that the relation between the unequal parts does not actually exist. It is merely comparative. "Exploitation" highlights the reality of the relation. It is extractive. An umbilical cord attaches the two "unequal" parts. It is not just that one party has more than the other. It is that one has more because it takes from the other and makes it less than it is. This is of major importance to us, given that both the domination of nature and colonial domination are exploitative/ extractive relations of this type: the thriving of the human and the thriving of the colonist happens through extractions from nature and the life-world of the colonized. Racial domination, then, resembles the process of dominating natural otherness, and also resembles a particular mode of extraction.

But it is important to say what this "particular mode" is. Not going beyond a comparison between bio- and necropolitics might stop us from highlighting exploitation, but thinking exploitation without necropolitics also stops us from capturing something crucial. The divide between biopolitics and necropolitics is not the divide between the life-world of the exploiter and the life-world of the exploited. It is the divide between spaces where people and animals are exploited within limits, and spaces where people and animals are exploited at whim without any formal restraint, up to the point where they lose the ability to regenerate themselves or (what is another way of putting it) are exploited to death. We can usefully speak of differential modes and intensities of extractive governmentality. It is this differential that a Marxist theory of capitalist accumulation can help us understand, providing us at the same time with a way of going from noting the similarity between colonial racism and the domination/ exploitation of nature to exploring a common generative mechanism behind the two.

Primitive accumulation

In looking at biopolitics and necropolitics as two modes of exploitative governmentality we need to remember that these happen concurrently, not in an either/ or fashion. They give rise to two social realities that coexist in a variety of ways within social space. Often they are directed at different groups of people or objects in different spaces, sometimes in the same space, but sometimes also they are directed at different dimensions of the same governed subject. That is, biopolitics and necropolitics constitute together a complex governmental assemblage defined by the the way the two are combined, dosed, and administered to a variety of groups. If they are not always (as Rosi Bradiotti argues) two sides of the same coin (2013: 122), they are at least coins issued by the same bank. The same governmental assemblage mutates to promote egalitarian citizenship, respect of law, democratic justice, "the good life," and regulated and measured exploitation in one instance, and violence, dispossession, discrimination, repression, and unchecked exploitation on the other. Likewise, the same governmental assemblage that promotes the protection of forests, parks, and wildlife in one instance facilitates a devalorization and an excessive economic instrumentalization of nature in the form of destructive extraction of resources and agribusinesses in the other. One of the most compelling theories we have to explain such fluctuations remains Marx's (1976) conception of primitive accumulation. This theory is of particular value to us, as it highlights the way capitalism works as an actual integrator of racial and ecologically destructive exploitation.

In explaining primitive accumulation, Marx begins by dismissing the classical economic fairy tale about the origin of wealth inequality: the idea that it all begins when some people accumulate wealth because they are "diligent, intelligent and above all frugal" (1976: 873), while others do not because they are lazy and decadent. Wealth was never accumulated in this fashion, Marx tells us. Instead, "in actual history it is notorious that conquest, enslavement, robbery, murder, briefly force, play the great part.... The methods of primitive accumulation are anything but idyllic" (p. 874).

The notion conveys both a sense of "unchecked and savage accumulation" and a sense of an "initial" or "original" form of accumulation which is developed in classical political economy, a historical moment that explains how capitalism began but which belongs wholly to the past, and has no relevance to what capitalism has become. Taking "primitive" to only mean "initial" has been academically discredited but it is nonetheless replayed every day in settler-colonial societies such as Australia, the United States, and Canada. Such societies continuously debate whether colonization was a historical event and, if it was, whether indigenous people should "get over it," as they are often told they should by conservatives. Or is it an ongoing process that continues to this very day? The interpenetration of the historical and the structural is more obvious in a younger settler-colonial society such as Israel, where it is impossible to know where the savage accumulation of wealth through violent appropriation begins and ends.

Early Marxist writing on imperialism quickly highlighted the permanent and cyclical nature of primitive accumulation. It was argued that, because the rate of profit tends to fall, capitalist societies have to continuously move anew to plunder land and resort to "savage" modes of creating capital. They were like those African tribal villagers who felt the need to regularly go back to the bush as a mode of rejuvenating themselves. Later Marxist theorists such as Samir Amin, Andre Gunder Frank, Martin Legassick, and Harold Wolpe continued to find in primitive accumulation an attractive explanation of how capitalism is linked to an ongoing colonial plunder well after "the end of colonialism" (see also

Arrighi 2009). Because the need for this plunder recurs, it quickly became clear that all colonialism, not only the settler-colonial process, is "a structure, not an event" (Wolfe 2006: 388). Capitalist nations must always fluctuate between civilizing themselves and allowing "lawful accumulation" to prevail such that they can foreclose the plunder, pillage, slavery, and genocide that generate their accumulated wealth. Yet they also must continually find places, within them or outside of them, where the rule of force can outweigh the rule of law so as to allow them to plunder, enslave, and rob once again. It is no different from hunting in this regard. In a hunting society, one cannot only hunt once, skin and cut the meat and make it neatly appear at the table divorced from the process of its production. One has to continuously do it. We can see here in this need for excessive exploitation the social roots of the need to fluctuate between biopower and the fostering of life on one hand, and necropolitics on the other: it is not just that we have the ecologically suicidal, and colonially violent, illegal and discriminatory society on one hand and the ecologically sustainable, lawful, egalitarian and democratic society on the other. It is that the former has to be continually revisited as the material condition of possibility of the latter.

In his analysis of the history of piracy Amedeo Policante (2015) sees in the pirates' plunder one of the
necessary roots of accumulated capital. He argues for
the existence of a "line" between piracy and legal property whereby the aim is always to plunder and cross the
line. In this sense capitalism always contains these two
realities separated by a line which works as a "laundering" mechanism transforming what is illegally obtained
in one into legal property in the other. Pehaps the
French word for laundering, "*blanchissage*," works even
better for us here since it literally means "whitening."
Indeed we can say that capitalism has an inbuilt "whitening" mechanism. This helps highlight the role of
race and racism in the process of illegally appropriating
"black economies" and transforming them into "white"
legally held property. Agamben's "state of exception" as
a framework for legitmizing and legalizing the continued need for illegal appropriation is crucial here. As
Policante rightly concludes, "the state of exception is
the legal form in which primitive accumulation can be
realized" (2015: 60). He explains,

> Plunder in the colonial world took place in a state
> of exception that negated, once and for all, the Uni-
> versality of Christian international law; however, the

wealth arriving in the ports of Europe would be soon registered and recognized as legitimate property. Plunder, piracy, theft – what originally had been a violent rupture – systematically becomes legal property. Beyond the line, all property is contested and yet, as Marie de Medici wrote to the Spanish King, "whoever proves the stronger shall be taken for the lord."... The stronger freebooter "beyond the line" becomes the legitimate holder of property on this side of the line. (pp. 59–60)

Consequently, we can say that primitive accumulation provides the structural basis of an apartheid-like tendency, a tendency to have at least two very different types of development and two different types of capitalist sociality at the very core of exploitative capitalism. One is defined by a civilized, cosmopolitan, state-regulated, lawful, welfare-supported, ecologically concerned exploitation. The other is defined by a savage, anarchic capitalism, spatially or socially peripheral to the cosmopolitan center (this center-periphery logic can be international or intra-national, it can even be intra-urban between two forms of social inhabitance of the same cosmopolitan city), and dominated by unchecked exploitation, theft, and pillage. One is regulated with a policing logic. The other is a space of war.

Primitive accumulation, overexploitation, and overoscillation

The idea that capitalist societies produce and need savage spaces and zones such as those described above reinforces the idea that the violence and discrimination that characterizes racism is not an aberration of democratic societies but a permanent feature associated with the government and exploitation of colonized spaces. But the logic of primitive accumulation tells us more than that two different spaces exist: it tells us that the civilized space of legality and democracy is *dependent* on the racist colonial space of unregulated accumulation for its existence, sustenance, and regeneration. Inside settler-colonial societies the fluctuation between the two realities is often conceived as a structural synchronic relation: the space of primitive accumulation is analyzed as "underlying" civilized cosmopolitan space. This is in line with Wolfe's characterization of colonialism as "structure not event" noted above. But the fluctuation has also been imagined diachronically as a kind of temporal circular logic. In this analytical vision capitalist society keeps alternating between a time of savage accumulation and a time of civilized sociality, for example, in the way people refer the rise of the welfare state in post-WW2 Europe as opposed to its current systematic dismantling. It seems to me, however, that both reflect an important dimension of the relation and they need to be thought together rather than in opposition. It is in this spirit that I will refer to this fluctuation as an oscillation that can be imagined both diachronically and synchronically.

Primitive accumulation, then, helps us delineate a state of permanent oscillation between two coexisting states of the social, the civilized and the uncivilized, as the very definition of capitalist social normality. The logic of the "tendency of the rate of profit to fall" that propels this normal oscillation also offers an explanation of how it can degenerate into crisis. Normality as a kind of acceptable and functional oscillation between regulated and unchecked exploitation, is predicated on the capacity of the oscillation to offset the tendency of the rate of profit to fall. If it doesn't, capitalist accumulation initiates a more intense drive toward the exploitation of human and natural resources. This often leads to both an overreaction from the exploited and a blurring of the "line"/border between bio- and necropolitics. As such, it is a chaotic oscillation that characterizes capitalist crisis.

There are many theorists who have seen a source of the ecological crisis in a similar excessive need for unchecked capitalist exploitation of natural resources. Capitalism in this savage state not only conquers land and exploits people but also exhausts what it delineates as nature, from soil (Clark and Bellamy 2009) to mineral and vegetable resources (Backhouse 2015). It does so by endless cycles of mindless unregulated extractions and appropriations (Moore 2011). This ontologically excessive drive for profitability leads Justin McBrien to argue that today "the accumulation of capital is the accumulation of potential extinction – a potential increasingly activated in recent decades" (2016: 116). Thus in his view, rather than the Anthropocene, capitalism leads us into the age of the Necrocene.

This "necrocinic" crisis can then be defined by an intensification of the capitalist drive toward overexploitation noted above. This tallies with earth systems scientists' arguments that the exploitation of the environment has led us to transgress "core boundaries" involving biodiversity loss and species extinction (Eckersley 2015). Transgressing core boundaries is, to use Spinozist language, akin to a "conatic transgression," a qualitative shift where what is extracted from the colonized is not merely "some of it" but its very capacity to "strive to persevere in its own being" (Spinoza 2005: III, 6). In a more anthropological spirit, we can say that it is a transgression on what we "keep" in the process of keeping-while-giving, an alienation of what should be inalienable (Weiner 1992).

It is a similar sense of transgression and overexploitation that pervades the experience of colonialism in the Middle East. For intimate observers of the region, it is hard to miss the necropolitical logic of unchecked accumulation in the process of "oil plunder" that has propelled the murderous invasion of Iraq and the destruction of Libya. This is not a conspiracy theory positing secret omnipotent plots, though it is equally naive to believe that plots do not exist. It is more often than not autonomous internal contradictions that create openings and possibilities for the lurking forces of colonial capitalism, which unleash dynamics that are independent of the initial cause of destabilization. The same can be said about Syria and Egypt, with their entirely different outcomes. In much the same way, it is impossible not to see the relentless forces of savage accumulation in the free rein given to the land-conquering Zionist settlers in Palestine. But this colonial overdrive is not specific only to the Middle East. In Australia as well, we witness a "free for all to see" connection between the endless desire of a powerful but troubled mining sector and the moves to further expropriate land from indigenous people and shrink their spaces of autonomy. A similar process has been occurring in the United States, leading more recently to the highly mediatized confrontation at the Standing Rock Sioux reservation.

In all of the above places, the "line" or border, national or symbolic, that orders "lawful accumulation" comes undone in the drive toward "unchecked accumulation," creating chaos in the very spaces that were previously ordered by capital itself. Even one of the most enduring legacies of the "classic" era of colonialism – the national border – is eroding. Colonially drawn and constructed, borders have served to reproduce the colonial relation well after the decline of the era of classical colonialism. They have worked as a means of dividing up and regulating the exploitation of resources as well as a means of containing and positioning colonized people in specific places. And while many people in the West are worried about their state's ability to protect the "sanctity" of the order of national borders, these "white worriers" show little awareness that their state has contributed to undoing that order elsewhere. The current destabilization of the Sykes-Picot borders in the Middle East and the flow of refugees from that region are prime examples of this breakdown.

Overexploitation signals a slow diffusion of the logic of primitive accumulation throughout social space. It creates a disturbance in the rate and tempo of oscillation between bio- and necropolitics and a certain lack of clarity as to the whereabouts of the line separating primitive from civilized accumulation. In "ordinary" times, each "white" civilized society aims to completely shield itself from the spaces of savagery that constitute its conditions of possibility. These colonial, neocolonial, and settler-colonial societies are all affected by a spatial politics aimed at managing relations between the spaces fostering the colonial good life inhabited by the colonizers (be it in Tel Aviv, Sydney, Paris, or New York) and the spaces of unchecked accumulation, offering more land to occupy or more resources to plunder. First and foremost is the line separating cosmopolitan goodness from colonial savagery, which aims to ensure that the umbilical cord of exploitation is not perceived to link them. It aims to ensure that its citizens' experience of the "good life" is not perturbed by the experience of the savagery needed for this good life to be experienced.

One of the key differences between the neocolonial Western nations and settler-colonial societies is their ability to foreclose this savage reality. In this context foreclosure is equivalent to a civilizational process (Elias 1994): the more a colonizing nation can shield its citizens from the savage realities that underpin it, carving out spaces where they are not exposed to the colonial conditions of their good life, the more civilized it appears. This has created a continuum rather than a sharp division, wherein the United States, Australia, Canada, and New Zealand sit between Europe and what is the ultimate settler-colonial society today, Israel.

At the same time, each colonial society tries to disavow the savage grounds on which it rejuvenates itself by distancing itself from those who have to be more openly savage than they are. The English and the French, both as governments and as peoples, tend to distance themselves from what they see as the outrageous Australian behavior toward indigenous people. In contrast, Australians – and North Americans – try to portray themselves as considerably more civilized than Israelis (whether or not they "support" them). But in much the same way, the citizens of Israel "proper" (which, from Policante's perspective, is itself the legalized result of a previous round of savage accumulation and plunder) look down on the inhabitants of illegal Jewish settlements in Palestine and aim to distance themselves from their "savagery." Yet, in the era of the permanent expansion of primitive accumulation, the line between the cosmopolitan centre and its savage periphery is blurred and the Zionist settlements become less and less of an anomaly: far more than the concentration camps highlighted by Agamben which are not governed by the logic of theft of land and resources, their savagery and in-your-face colonizing *raison d'être* are metonymic of our era.

As the above amply demonstrates, the analytic assemblage delineated by the declining rate of profit and primitive accumulation offers us an important insight into the way capitalist exploitation works as the same structuring principle behind ecological and colonial domination, as well as a key determinant of the structures and the forms taken by the crises in both domains. We can easily say that colonial racist exploitation reproduces and legitimates the very wild, unchecked, and inhumane capitalism that governs the overexploitation of nature. Nonetheless capitalist exploitation, for all that it explains, keeps unexplained at least one large facet of both crises: their affective dimension. Going through some of the more important expressions of these affective realities will guide us to what I consider a more fundamental mechanism behind racial and ecological exploitation. The first of these affective states is the sense of being besieged, generated by both crises.

Fantasies of colonial reversal

"More Evidence Europe Is Under Siege by the Flow of Illegal Migrants from Africa": so reads the headline for a piece written by Ann Corcoran (2015), a member of Refugee Resettlement Watch, an organization that purports to provide the latest news on "the invasion of Europe." "Germany in a State of SIEGE," screams a UK Daily Mail headline. "Merkel was cheered when she opened the floodgates to migrants," the title continues in bold. "Now, with gangs of men roaming the streets and young German women being told to cover up, the mood's changing" (Reid 2015). But this association of refugees with a state of siege is not confined to ultraright organizations and tabloids. Hardly any newspaper – whether antagonistic to asylum seekers, such as the Australian *Daily Telegraph* (September 9, 2015), or sympathetic to their plight, such as the *Los Angeles Times* (August 6, 2015) – failed, at least occasionally, to refer to refugees in terms of "flows," "flood," and "waves." Meanwhile, the feeling of siege is reinforced by government actions, such as Australia's deployment of its navy to confront asylum seekers arriving by boat, the United States' erection of walls on its militarized border with Mexico, and European states such as Hungary laying out barbed wire along their borders.

"The European Migration Crisis: Europe Under Siege," goes the headline from the *American Interest*. "The migrant crisis is not an easy problem, and none of the answers are good," says the writer, who continues:

"But control over its borders is a necessary task of an effective state, and ultimately neither in Europe nor in the US will we be able to have generous and humane asylum and migration policies if voters lose faith in the will and the ability of the political leadership to do its clear duty and police the frontiers" (*American Interest* 2015).

It is particularly this sentiment – that the existing order of national borders is being tested by "flows" of people who seem intent on disregarding it – that is the prime source of this sense of besiegement. "Europe Under Siege?" asks Daniel Gros, director of the Brussels-based Center for European Policy Studies, somewhat less assertively. But his opening sentence leaves no room for doubt: "Many Europeans feel like their countries are under assault, as huge numbers of migrants flow across their borders" (Gros 2015). These media snippets serve as metonyms for a wider private and public cultural experience of feeling under siege.

At one level, we can argue that these fears have a long history. A feeling of being besieged by the very people whom one is actually colonizing is, paradoxically, part and parcel of the history of colonialism. Images of Asians, indigenous people, Arabs, and black people dominating, exploiting, and enslaving white Europeans abound in colonial cultural production. This is

so even at the height of the "classical" European colonial venture when colonialists were, relatively speaking,
most secure about their entitlements and their transnational supremacy. Stephen Arata (1996) has called these
stories of besiegement "narratives of reverse colonization." As he put it, discussing English literature,

> Moments like this recur with remarkable frequency in
> late-Victorian popular fiction, embedded in what can
> be called narratives of reverse colonization. In such
> narratives what has been represented as the "civilized"
> world is on the point of being overrun by "primitive"
> forces."...In each case a fearful reversal occurs: the
> colonizer finds himself in the position of the colonized,
> the exploiter is exploited, the victimizer victimized.
> The reversals are in turn linked to perceived problems
> – racial, moral, spiritual – within Great Britain itself.
> (1996: 255–6)

It is perhaps not surprising to see the theme of "reverse
colonization" reemerging at the very time when we
are witnessing an intensification of the Western colonial military interventionism in the Middle East that
began with the US invasion of Afghanistan in 2001.
According to a piece titled "Reverse Colonialism and
Its Cause," posted on the God's Kingdom Ministries
website, which addresses "serious Bible studies,"

Europe is in crisis. Refugees are rushing into Europe as a result of the wars in Syria, Afghanistan, Iraq, and even Pakistan. They say they want to have a better life and to live in peace. Those are the words of colonists who have no intention of returning to their country. From refugees, they have become immigrants, but large masses of immigrants who have no intention of assimilating into European culture are actually colonists. (God's Kingdom Ministries 2015)

Another web source describes a "first encounter" with the concept, displaying nonetheless an awareness of the actual exploitative dynamics of colonialism:

Today for the first time I heard the phrase "reverse colonialism." This refers to the observation that after centuries of Eurocentric cultures exploiting the rest of the world for their own economic gain, the rest of the world is – de facto – colonizing these first world countries back. England is filling up with people from India, France is becoming populated by Muslims who bring their own culture with them, and so on and so on, in Germany, Holland, and various other nations that have become wealthy over the centuries with the assistance of a somewhat exploitive world economic order. (Ken's Blog 2009)

Perhaps one of the classic literary expressions of these paranoid fantasies of reversal has been Michel Houellebecq's novel *Soumission* (2015). This novel imagines a peaceful and democratic Islamic takeover of France, which ends up generating a creepy sexist dictatorial social and moral order, and a whole new class of French "collaborators" that happily join in, lured by all kinds of equally creepy sexual and financial benefits.

The fact that we are talking about a novelist of the stature of Houellebecq should not deter us from thinking that he is capable of producing something of that paranoid order. As Arata points out, many a good novel belongs to the genre, from Stoker's *Dracula* to H. G. Well's *The War of the Worlds*. Nonetheless, and as he rightly observes: "Stoker's Count Dracula and Haggard's Ayesha frighten not least because their characteristic actions – appropriation and exploitation – uncannily reproduce those of the colonising Englishman" (Arata 1996: 108).

In *Soumission*, the Islam that ends up taking power is really an unnerving product of a genuinely colonial imagination seemingly unaware of the coloniality it is reversing. So is the novel's racialized sexism, characteristic of most forms of Islamophobia, which ends up producing subservient Islamic women en masse, all willing to stop working as soon as Islamic rule shows its face, all willing to become mothers, feeders, and whores without a whimper.

Interestingly for us, those same fantasies of reversal have also been part of the history of domesticating nature. Here too we have a long record of worrying that what we have dominated and exploited will rise against us and domesticate us in turn. In some instances, this has taken the form of an imaginary wherein the plant and animal world "reclaims" the human domain of the built environment. Such fantasies also feed on images of destroyed or deindustrialized cities, such as Beirut during the civil war or Detroit taken over by plants. This is how the director of a recent documentary about Detroit describes the urban spaces that he has filmed:

> The vast, rusting hulks of abandoned car plants, some of the largest structures ever built and far too expensive to pull down, beached amid a shining sea of grass. The blackened corpses of hundreds of burned-out houses, pulled back to earth by the green tentacles of nature. (Temple 2010)

Sometimes the fantasies of reversal are identical to the colonial fantasies of reversal, imagining worlds where we are ruled by the very animals we have dominated. *Planet of the Apes* is but the most obvious example of this imaginary circulating in popular culture. In his novel *2007* the Australian naturalist Robyn Williams (2001) also takes us into a world where a more generalized process of reversal is under way. Phillip Armstrong gives a good summary of the novel:

> The story opens with simultaneous acts of civil disobedience on the part of various species: forty baleen whales sink a Japanese whaling vessel, a mile-wide flock of pelicans occupies Heathrow airport, hundreds of cows invade Melbourne's Tullamarine Freeway, pythons attack a line of bulldozers poised to raze Amazon rainforest, foxes mass against a Buckinghamshire hunt. The world's pets, racehorses, farmed and zoo animals break out of their confines and, along with rebellious wild species, occupy the parks and green belts of the world's cities, which are soon densely packed with animal refugees....
>
> *2007* reflects a growing anxiety that the relentless expansion of modernity will inevitably create a reaction; that there is no place left for human enterprise to occupy without coming into punishing conflict with the natural world. (Armstrong 2008: 171)

These fantasies of reversal raise the same questions as those raised in relation to their colonial equivalents: while the fears are real, to what extent is the imagined threat itself real? And while, as with colonial fantasies, these imagined reversals have existed ever since we began dominating nature, how do we understand them today in the face of global warming? Is there no difference in the fears embodied in Alfred Hitchcock's *The Birds,* based on a 1952 novel, and those embodied in the film *The Day after Tomorrow*, which came more than fifty years later?

With the intensification of the ecological crisis and its effects, particularly in the form of global warming, our capacity to imagine natural phenomena as having an agency that can be turned against us has increased, as have our fears and the actual threat of natural catastrophe. In Armstrong's text, there is still a reluctance to think that nature is actually acting back. As he put it in the last sentence quoted above: "there is growing anxiety that the relentless expansion of modernity will inevitably create a reaction." Despite the "inevitably," there is a modernist hope that the inevitable will not happen. One of the distinguishing features of the works of Bruno Latour today is to argue that one is no longer dealing with mere possibilities: the inevitable has happened and the earth is already reacting. In a recent work, *Face à Gaïa*, Latour (2015) opens with the figure of a dancer who metonymically conjures the state of being beseiged. She continuously looks behind her, worried about something ominous that is about to happen. This something ominous is for Latour the figure of Gaia, the inevitable figure born out of the earth's overexploitation. Like the figure of the Muslim born out of colonial overexploitation, it is out of control and it is kicking back: an ungovernable, overexploited entity running amok. This takes us to the second affective domain worth highlighting: the affects of ungovernability and the dreams of exterminability associated with them.

Ungovernability and exterminability

We have already seen in the previous chapter how the already affectively charged category of the wolf articulates itself to the category of the ungovernable. This kind of affect and its imaginaries already put the ungovernable outside of what the logic of capitalist accumulation can explain by itself. To understand what kind of formation can generate such wolfish affect we need to dwell further on the peculiarities of the ungovernable as a category.

On the face of it, this is reasonably simple. The politics of the ungovernable emerge when a governmental system, a governmental culture, or any other governmental assemblage faces a particular object, or a general environment or situation it wishes to control and yet finds itself unable to do so. It is unable to do so either because there is a crisis of method – it is unable to govern in any of the particular ways it wishes to proceed – or because the object itself is immune to any available technique of governmentality. But to fully understand the affect associated with ungovernability we need to be clear that the ungovernable is not simply something outside the space of a governmental apparatus, or what we might call "the ungoverned." There are many things around us that are ungoverned: things that are beyond our reach, our interest and/or our control. These are functions of both the limitations of our governmental capacity, and the accessibility and complexity of certain aspects of our environment. That can go from micro-scopic processes occurring inside various organisms to the movement of the planets, and to things we are not even aware exist around us.

Nor should the ungovernable be mistaken for the "not-governed." Unlike the ungoverned above, the not-governed is accessible to a governmental apparatus. It can be something "not governed by us": that is, we accept that it is part of another governmentality, like something or someone subjected to the laws of a nation other than ours. This could be the nation next door or the grounds of a foreign embassy within our nation. It can also be something new that a governmental apparatus has classified and considered as "not-governed-yet": that is, its becoming governed in the future is considered desirable and potentially feasible. The governmental expectation is that it is only a matter of time before the not-governed-yet becomes governed. The not-governed can also take other forms. Rather than be a "not-governed-yet" it can be something considered "better-left-not-governed." That is, not everything that is not-governed is so because it cannot be governed. There are many things we have an interest in letting be, and in keeping not governed. As Foucault (2008) points out in the introduction to his 1978–79 Collège de France lectures, the art of modern government comes hand in hand with the recognition of the limitation of government and, most important, the capacity for self-limitation. Foucault exemplifies this governmental sense of self-limitation with the comment by the English statesman Walpole: "Let sleeping dogs lie" (p. 10). Here, Foucault explains, it is not the question of governmental power that is at stake but the question of governmental know-how. A government that does not know when to "let sleeping dogs lie" and wants to govern everything is paranoid, clumsy, and inefficient.

The ungovernable is of a different order, and needs to be clearly distinguished from the above. To begin with, we can say that the ungovernable is of the order of what we desire to govern and subject to our power, but are continuously and repeatedly unable to do so. That is, the ungovernable is something we have originally classified as what we have referred to above as "not-governed-yet" – which also implies that we consider it as an "ought-to-be-governed" – but find ourselves, and one must insist on this point, find ourselves repeatedly unable to govern. It is because of this repetitive nature of the encounter that we start classifying something as ungovernable. Thus we can say that the ungovernable denotes a certain history. Individuals, groups, animals, plants, things or social processes cannot be deemed by a governmental assemblage as "ungovernable" on the basis of a single encounter. This makes "the ungovernable" as a classification and an experience paradoxical in that it indicates on one hand an inability of a governmental force to relate to it and yet it also implies a historically acquired familiarity: it denotes a relation paradoxically marked by a certain intimate lack of relationality, a relating to something through a recognition of the permanent inability to relate to it. It conjures the analytical images of the Freudian uncanny that have often been widely mobilized to understand "the other" (see in particular Kristeva 1982).

The imaginary figure of the Muslim as wolf exemplifies the figure of this frustrating object, around us but escaping us, in the realm of our governable sphere and yet ungovernable: the other that continuously threatens our desire to feel in control of our environment. It generates in us a very particular set of affects associated with the threat of loss of sovereignty. The more an object's ungovernability endures, the more it haunts and threatens us. It puts us face to face with our vulnerability and the limitations of our sovereign power and makes us desire to govern it even more intensely while at the same time fearing that it will be forever ungovernable. It is such permanency that animates, in those who persist in desiring to govern, the exterminatory fantasies that become integral to the experience of ungovernability and that are directed toward the Muslim today.

Generalized domestication as a fantasy of viability

As noted above, the fact that the colonial and the ecological crises generate similar affective tendencies associated with a sense of loss of control and sovereignty, point to them sharing more than just an objectivist "logic of capitalist exploitation." The emphasis on environmental control, sovereignty, fear of reversal, all experienced as something not merely "functional" but as something in which the controlling/exploiting/governing subject has an affective interest, something where the subject's very viability is on the line, so much so that they harbor exterminatory tendencies toward those who disrupt their desires and their viability, puts us face to face with generalized domestication as a fantasy of viability.

The quickest way into understanding generalized domestication is to begin by associating it with instrumental reason. This is a reductionist understanding of it but a useful entry point. Indeed instrumental reason is often conceived as the least affective of all forms of reason. Implicitly informed by the way it has been captured by Spinoza's notion of striving, Heidegger's "home-building," and Nietzsche's "will to power," my aim is to show that it is part of a highly affective mode of existing in the world.

Generalized domestication is a mode of existence. That is, it is a phenomenologically understood mode of being. It can also be referred to as a mode of enmeshment, a mode of inhabitance, and a mode of deploying oneself in the world. These are all aimed at highlighting a mode of relating. It is, however, a mode of relating to the world that, in the process of relating, creates the very world it is relating to. (For more on this mode of conceiving the production of reality see Jakob von Uexkull 2010, through Deleuze 1988, to Viveiros de Castro 2014.) What best defines this mode of existence is that it struggles to create a world where the most salient quality of everything that comes into existence is that it "exists for" something. The variety of aesthetic, economic, affective, and other ways in which things can be seen as "existing for" a domesticating subject are innumerable, but it is this instrumentalization in its most general sense that defines generalized domestication as a mode of being. It is crucial, however, to understand that, in its striving, generalized domestication does not offer us "a point of view" on an already existing world. Rather, the relation or the form of enmeshment as it unfolds brings into existence an actual world where a key dimension of this world is its "existing for" a domesticating subject. It is this production of a meaningful reality that allows us to see in it "a fantasy of viability." This has to be understood in a loose psychoanalytic sense. This is not the ordinary sense of "fantasy" as something that is a product of the imagination that is detached from reality. Rather it means that it is a staging of the self that allows the self to come into existence: the creation of a meaningful space whereby, at the same time, one gives meaning to one's own life as a life worth living. It is a fantasy where existential viability depends on one's viability as a domesticator. This is why the ungovernable is experienced not merely as a technical problem but an existential threat (as Israelis call every political force that seems outside their control). In this sense, it is a continuous replaying, at the level of everyday life, of the grand instrumentalist biblical tale of creation:

> the fear of you and the dread of you shall be upon every beast of the earth, and upon every fowl of the air, upon all that moveth upon the earth, and upon the fishes of the sea; into your hand are they delivered. Every moving thing that liveth shall be meat for you. (King James Version, Genesis 9: 2–3)

However, neither this staging of the viable self nor this instrumentalization gives generalized domestication its specificity like the peculiar way it frames the process of home-building. As we will argue in the next chapter it is this mode of home-building that is infused in the practices aimed at both natural and colonial/racial domination.

THREE

The elementary structures of generalized domestication

In the previous chapter I argued that despite its account of the way in which the racial/colonial crisis and the ecological crisis are related, a capitalist theory of exploitation and primitive accumulation does not allow us to understand the affective dimensions that mark the way both crises are experienced. The logic of capitalist exploitation can tell us why the two modes of domination are interrelated in a macro processual sense, but it cannot tell us at the level of everyday experience how racist practices and the practices of environmental domination partake in each other's realities. In this chapter, I want to go into more detail about the fantasy of viability that I have called "generalized domestication" to show how this reproductive complicity is generated.

As already noted, theorists of animal domestication have often recognized the particular similarity between the practices of domestication and certain forms of interhuman relations. In his history of domesticated animals, Zeuner points out that

> it is necessary to remember that the social relation called "domestication" is by no means restricted to man and his animal subordinates. Man has applied the same practice to members of his own species, though in this case it is usually called slavery, unless a more euphemistic word is used. (Zeuner 1963: 36)

At the same time, the notion of "domestication" is loosely and freely used in social analysis to describe any general process of coopting or taming a potentially dangerous or alien social force (such as *The Domestication of Women* [Rogers 1980]). While following a similar path, the conception of generalized domestication that I am proposing here is more precise in what it entails. Keith Thomas remarks in *Man and the Natural World* that the domestication of nature is "the archetypal pattern for other kinds of social subordination" (Thomas 1983: 46). My interest is partly grounded in developing what this archetypal pattern entails and how colonial racisms such as Islamophobia follow such an "archetypal pattern." As should be clear by now, however, my aim is not to merely indicate that racism is "similar to" the processes of domesticating nature.

Generalized domestication, as I want to develop the concept, is a mode of inhabiting the world through dominating it for the purpose of making it yield value: material or symbolic forms of sustenance, comfort, aesthetic pleasure, and so on. There is no intention to give primacy to the domestication of nature and particularly animals, as the most "fundamental" or "real" form of domestication, that others look like or do not look like. Jean-Pierre Digard remarks in his *L'Homme et les animaux domestiques* that there is nothing in the concept "domestication" that links it necessarily to the practices of dominating the natural world. He reminds us that "the adjective domesticated applied to animals…does not appear until the fourteenth century" (Digard 1990: 24, my translation). Nonetheless, the domestication of nature has been and remains the instance where such a mode of existence has been spoken about and detailed in the most explicit way. Thus references to the way it has been conceived do help us understand the more general concept we are aiming for. Furthermore, while generalized domestication is not only about the domestication of this or that species but the domestication of one's whole environment, what is true of the logic of domestication in relation to one species is also true in relation to the environment as a totality.

From domestication of nature to generalized domestication

A zoological definition of domestication, such as the one offered by Bökönyi below, is considered by many to be an "orthodox" way of describing the practice:

> The essence of domestication is the capture and taming by man of animals of a species with particular behavioural characteristics, their removal from their natural living area and breeding community, and their maintenance under controlled breeding conditions for mutual benefits. (Bökönyi 1989: 22)

Another definition, of equal importance, is given by Pierre Ducos, who offers a synchronic look at domestication and stresses the status of the domesticated as an object of usage and value to the domesticator. He argues that

> Domestication can be said to exist when living animals are integrated as objects into the socio-economic organization of the human group, in the sense that, while living, those animals are objects of ownership, inheritance, exchange, trade, etc. As are the other

objects (or persons) with which human groups have something to do. (Ducos 1989: 24)

One element of importance in the first definition is the diachronic depiction of the consecutive moves necessary to obtain a domesticated species. Any domesticated species has to be, historically speaking, captured and tamed before it can become domesticated. Nonetheless, the aim is to make it reproduce itself in captivity. That is, in general a domesticated species is one that no longer needs to be captured and tamed, though this capturing and taming have been the essential historical condition of possibility of its domestication.

At another level, however, this first position reproduces one of the most enduring ideas that has accompanied domestication, and that is the idea of "mutual benefit." This idea can even be found in Aristotle, who argued that "all tame animals are better off when they are ruled by man; for then they are preserved" (Aristotle 2014: 1990). Keith Thomas explains how "in the eighteenth century it was widely urged that domestication was good for animals: it civilized them and increased their numbers: 'we multiply life, sensation and enjoyment'" (Thomas 1983: 20). In a similar vein, some of the ways domestication is popularly conceived, in opposition to *predation*, highlight its supposed nature as a symbiotic relation when compared to the latter. As is well-known, this idea of "mutual benefit" seeps into the discourses surrounding slavery.

At one level, the idea of mutual benefit is somewhat difficult to argue, since the animal that is "benefiting" from being in a relation of domestication is not the same animal as the one that existed before entering the relation. As Bökönyi (1989: 25) herself points out: "The result of domestication is that the domesticated animal first culturally and later morphologically differs from its wild form." Class relations, even when unequal, can also work this way. For instance, the capitalist-worker relation can be said to be of mutual benefit to both. But this presupposes and normalizes the transformation of people into laborers. The same can be said of slave relations, of patriarchal relations, and of colonizer-colonized relations.

It is nonetheless important to recognize that the idea of "mutual benefit" is not some kind of mystification of the truth that aims to "overshadow" (Digard 1990: 102) the relation of power in which domestication is grounded. While there is clearly an element of truth in this, it is an experiential fact that some relations, despite being grounded in domination and exploitation, are nonetheless experienced as "mutually beneficial."

Experiencing domestication as being of mutual benefit does not stop it being a relation of power structurally grounded in the initial practices of capture and taming that initiated it. Seeing "mutual benefit," with a kind of "Marxist" eye, as so much mist hiding a real "reality," ends up itself actually hiding what is a crucial reality of domestication: it is a relation of domination that aims to be lived as a relation of nondomination. Its appearance is not what hides its essence but is part of its essence.

This is why understanding domestication purely in terms of its objectifying, instrumental logic, as the second definition by Ducos allows us to do, reveals but also conceals one of its important affective dimensions, conveyed by the very meaning of the word. Domestication is after all a struggle to make things partake in the making of one's home. It is a struggle to create homely spaces or, to put it more existentially, a struggle to be "at home in the world." Yet, paradoxically, it is also a mode of domination, control, extraction, and exploitation.

Emile Benveniste's tracing of the etymological roots of the word "domestication" helps us understand this paradox (Benveniste 1973: 243). Most people acknowledge at least implicitly the links between domestication and the notion of the *domus*, and as such recognize it to be a practice of making "homely" or of "bringing into the home." Benveniste details this link but also highlights the fact that *domus* itself shares its roots with *dominus*. Domestication is not just any kind of homeliness, it is a homeliness obtained through domination. It is precisely here that its most important feature, as a fantasy of viability, resides. While the aggressive affects associated with "domination" are often thought of in opposition to the gentle and cuddly affects associated with the home, domestication offers us an understanding of one of the crucial modes in which we exist and where these two affects are inseparably intertwined. Generalized domestication is the fantasy whereby we make our existence viable by seeking homeliness through aggression and domination. This is why one of the essential forms of labor that define it is the labor of managing the relation between the spaces of aggression and domination that are constitutive of it and the homely, cozy, and warm spaces that are equally entangled with them.

In an interesting way, generalized domestication offers at a micro level, and in a kind of cellular form, the same drama we saw played out between primitive accumulation and civilized accumulation, which allows us to understand the former as the illegality behind (both historically and structurally) legality, the theft behind property, the violence behind peace, and the discrimination behind democracy. It is in much the same way that the relation of "capture and taming" is both the historical and structural condition of existence of the relation of "mutual benefit" that constitutes domestication. While not all relations of domestication can be experienced as homely and cozy relations of "mutual benefit," it is the function of domestication to ensure that the homely space is protected from exposure to those practices that can't. Domestication, therefore, aims not only at positioning things in the proper way to extract value from them, but also at ensuring that the value extracted is delivered in a homely way. This is why it is perhaps best exemplified by the most extremely patriarchal forms of mothering, insofar as "being mothered" is imagined as the ultimate mode of being "at home." In its patriarchal form, however, this mothering is impossible without it being grounded in the patriarchal relation of domination that has to nonetheless disappear from view for "mothering" to be truly "motherly."

It is within this homely patriarchal fantasy of governmental control, extraction and exploitation that one finds that the person aiming to dominate and utilize "nature" in order to feel at home in the world and the person wanting to racialize, dominate, and control the "Muslim other" in order to feel at home in their nation are at a fundamental level engaging in one and the same practice. In what follows I will examine the key features that make up this generalized conception of domesticating practices. I will begin by outlining what I will call the elementary features and will then proceed to say in what way they are elementary. These are: the process of occupation, the process of polarization between domesticator and the domesticated, and the practices of spatialized extraction.

The process of occupation

Domestication is first and foremost a mode of inhabiting the world by occupying it. Occupation here is meant in the settler-colonial sense. Indeed, from an inter-species perspective, every human occupation is an act of settler colonialism since one occupies a space that is always already occupied by other domesticators, whether insects, animals, plants or trees. Each of these inhabits the world with some degree of instrumentalization too: a tree spreads itself above and below the ground in its struggle to extract nutrition, sun, and so on. Ants also organize and transform their surroundings in a specific way. What defines human generalized domestication is the act of occupying a space by declaring one's own interest as its primary organizing principle. As such it relates to prior occupiers of the same space according to how their being can be harnessed to the advancement of our own being. What comes in the way is excluded or exterminated. This is true of people having a picnic as well as those organizing a farm, those building a house as well as those building a nation.

Herzl's imagination of the future building of the state of Israel provides us with a good example of a fusion between national and natural domestication.

> If we wish to found a State today, we shall not do it in the way which would have been the only possible one a thousand years ago. It is foolish to revert to old stages of civilization, as many Zionists would like to do. Supposing, for example, we were obliged to clear a country of wild beasts, we should not set about the business in the fashion of Europeans of the fifth century. We should not take spear and lance and go about singly in pursuit of bears; we should organize a large and active hunting party, drive the animals together, and throw a melinite bomb in their midst. (Herzl 1972 [1896]: 28–29)

What is particularly interesting above is that Herzl is not an abstract theoretician of the nation. He is

speaking as a nationalist and as someone who is thinking practically about instituting a nation. That is, he
is a nation-builder, or better still for our purposes, the
builder of a national "home." At the same time when
he writes this he still does not know where the future
Israel will be. So he does not have any particular "bear"
in mind. Yet as a domesticator he knows that there is no
occupation without a killing of bears, as it were. As such
he provides us with an example of the general domesticating imaginary behind nationalism in its purest form.

The "bear" or the wolf is always haunting this fantasy
of national homeliness. As such, a domesticated space
is a space that is, as Derrida (2008) says concerning the
space of sovereignty, a space that is always lacking. In
a formulation that is particularly suggestive in light of
how we analyzed the racialized images of the Muslim in
a previous chapter, Derrida argues that such a lacking
space where one experiences a *"manque de souverai-
neté"* is occupied by a wolfish figure of the other. It is
a furtively arriving wolf that is slowly coming (*à pas
de loup*), and yet is never there enough to fully exist
(*il n'y a pas de loup*) (Derrida 2008: 2). Nonetheless,
it is there enough to lurk around stopping us from
reaching our ultimately homely/motherly space. This,
in a sense, relays more evocatively what every farmer
knows and what every farming manual says, and that

is, that domestication is a never-ending process (Digard 1990: 172).

Indeed, the fact that domestication is an endless pursuit is one the key elements that qualifies it to play the role of a fantasy of viability. It sets the domesticating subject on a stage where one is forever aiming to maximize homeliness without ever reaching one's ultimate aim. There is always a wolfish something, Muslim or otherwise, who insists on having his/her own domesticating law, his/her own mode of being at home, in the space that I am occupying.

The process of polarization

One of the most common ways today of combating at a conceptual level the essentialism that is often the mark of relations of domination based on racism, speciesism and sexism is to highlight identities as "social constructions." Whether between humans and animals, men and women or whites and blacks, social/political critical analysts highlight that such differences are not "natural" and emphasize that they are constructed.

A while ago now some feminists reacted against thinking too much in terms of social construction. It would entail the portrayal of women's and men's bodies as if they were undifferentiated material that can be simply molded and shaped (constructed) to form the gendered person, male or female (Gatens 1996).

One of the dynamics emphasized by the logic of generalized domestication reconciles both of these tendencies. It does not oppose "construction of difference" to "intrinsic difference." Rather, it highlights the process of polarization of difference or, more exactly, the process of turning difference into a polarity. Thus, rather than emphasizing the way difference between humans and animals is socially constructed in opposition to a preexisting difference, domestication allows us to recognize that what matters most is not difference as such but that particular way of experiencing difference as a polarity. This polarization is a process driven by the domesticator that has an interest in the polarity. To speak of polarization as opposed to just difference is to speak of a difference where a force is aiming to evacuate each element of what makes it similar to the other.

The domestication of animals abounds with examples of attempts by humans not only to differentiate between humans and animals but to evacuate the human, as much as possible, from anything perceived as animal and vice versa. This is not the same as emphasizing the often mentioned process of conceptual differentiation whereby humans are defined in opposition to animals. It is rather to highlight an active process of bodily transformation. Norbert Elias's (1994) seminal work on the "civilization" process gives us a good example of the work done by humans to de-animalize themselves.

Marcel Mauss gives us an example of this labor of transformation in his techniques of the body when talking about the way he was "educated" to not walk like an animal with "his hands flapping wide open" (Mauss 1973: 72). Keith Thomas provides a history of how the human-animal differentiation was maintained through various interdictions. Some of those he listed were: "Don't smack your lips like a horse, he warned; don't swallow your meat without chewing, like a stork; don't gnaw the bones, like a dog; don't lick the dish, like a cat" (Thomas 1983: 37). Furthermore, as he points out,

> most people were taught to regard their bodily impulses as animal ones, needing to be subdued.... Lust in particular was synonymous with the animal condition, for the sexual connotations of such term as "brute," "bestial" and "beastly" were much stronger than they are today. (p. 38)

For Thomas, "Descartes had only pushed the European emphasis on the gulf between man and beast to its logical conclusion...man stood to animal as did heaven to earth, soul to body, culture to nature" (pp. 34–5).

Booker T. Washington in *Up from Slavery* (1963) relates how the exaggerated and genteel cleanliness of the slave-owners' establishments served to mark them off from the "animal-like" slaves, whose enforced filthiness (they were provided with no means to wash) served the joint function of marking and justifying their condition, and of linking them to animals.

Similar processes have been analyzed in sexist, racist, and colonial relations of domination. Albert Memmi (1965: 91) famously argued that "colonisation creates the colonised just as it...creates the coloniser." What is important for us is that it works at making polar opposites of them. In a wonderful work, *Making England Western*, Saree Makdisi shows that Britain's difference from its colony was labored on intensively to make the polarity Metropolis/colony a reality (Makdisi 2014). In much the same way and more in line with our preoccupation, we know that since the early stages of Western colonialism in the Middle East there has been a concerted effort at evacuating the long contribution of Islamic culture to the making of European modernity (see Mallette 2010).

The process of polarization clearly works through the valorization of a certain definition of "humanity" that

the domesticator aims to "evacuate" from the domesticated, be it "nature," a particular animal, blacks, or
Muslims. At the same time the domesticator aims at
increasingly acquiring the very thing it is evacuating
from the other. For example, as the human domesticator struggles to make the animal more and more animal
and less and less human, they also struggle to make
themselves less and less animal and more and more
human. The attainment of this "humanity" becomes a
project rather than something one is. This brings us to
the key fantasy element in the process of polarization:
The humanity of the domesticator, like their sovereignty, is something that is never complete but needs to
be continuously aspired for. It becomes a form of what
Bourdieu would call "cultural capital": something that
needs to be continuously struggled for both in terms
of its recognition and in terms of its accumulation. In
that sense even when they project themselves strategically to be "complete" there is never a domesticator, a
human, a white, a male person. It is always a trying-
to-be-domesticator, trying-to-be-human, trying-to-be-
white, trying-to-be-male person. This logic manifests
itself within Islamophobic discourse in the portrayal
of the Muslim as lacking some of the most fundamental attributes of what makes the modern human
being. The Islamophobe on the other hand is feverishly

trying to accumulate democracy, nonsexist behavior, tolerance, capacity for irreverent humor, anything that the Muslim is portrayed as having less and less of.

The spatial practices of extraction

The practices of shaping and managing the space that we occupy, that we differentiate and distinguish ourselves from, and that we exploit (as a totality and as a multiplicity of particular others) in order to make it yield "motherly/homely" value, remain the most important dimension of domestication. This involves both the eradication of that which can harm us, and the appropriation, positioning, and shaping of the being and mode of existence of whatever we find useful into a being and an existence for us.

Positioning involves not only positioning so as to facilitate extraction of value, but also positioning so as to minimize the effect of waste produced from the process of extraction. The process of exploitation can be broken into many instances such as appropriation, transformation, maintenance, consumption, waste management. All these practices involve particular spatial considerations as to where they can best happen.

The domestication of animals, for instance, involves claiming ownership through capture, the process of taming and molding of animal bodies (from clipping fur and cutting tails to the breeding of specific domesticated species), maintenance such as feeding and cleaning, the consumption of their labor power, of animal products such as milk or wool, or the consumption of the animal itself, and finally the management of the waste produced in all these processes of consumption. Domestication is not only concerned with single animal bodies. It is also about the spatial management (sequestration, caging, positioning, transporting, etc.) of animal populations (stocks). Consequently, a general theory of domestication has to encompass an account of power that emphasizes both the micro spatiality of constituting/modifying/exploiting bodies and the macro spatiality of population control.

This has clear affinities with Foucault's disciplinary and biopolitical conceptions of power (see Novek 2012). Domestication involves both the making of docile and productive bodies, and an "environmental" government of populations. It has a necropolitical dimension and a politics aimed at the fostering of the life of the domesticated. Indeed, it can be said that one of its primary aims is the forging of a docile body that "may be subjected, used, transformed and improved" (Foucault 1979: 136). It also has a particular affinity with Guillaume de la Perriere's conception of government in his treatise *Miroir de la Politique* (1567): "government is the right disposition of things arranged so as to lead to a convenient end" (in Foucault 1991: 93).

Nonetheless, it is important to recall here that we are not only talking about the form and practical principle behind formally recognized relations of power such as human-animal, or men-women, or colonizer-colonized relations. To say that domestication is a mode of inhabitance and enmeshment is to say that it is how we exist in the world, generally, even in the most ordinary of situations. That is, for example, it is also true of a man adjusting himself in his lounge chair so as to make it yield comfort, perhaps using a particular type and number of cushions to prop up his back, an overhead light to ensure good vision, setting out a footrest and maybe even a little table with a drink, while positioning a book on domestication in his lap so as to read it properly without tiring his hands or his brain. And who knows, sitting on that lounge chair might have involved an act of killing an ant or a mosquito that was in the way.

What is crucial about such a domesticating practice of environmental management is that it is on one hand based on an instrumentalist mode of classification that is reducible to questions of usage and harm. At the same time, these classifications are intimately associated with spatial positioning. Not only in the sense that a classification is done according to where something or someone is positioned, such as: a cockroach is "harmful" on my pillow but less harmful on the footpath. But more importantly in the sense of domestication itself being an active practice of positioning: classification happens as part and parcel of the process of positioning. We draw closer to us what we see as harmless and useful and we push away what we see as harmful and useless. To be sure, the useful/harmless–useless/harmful opposition operates more like an axis on which the domesticated object is positioned. Just as important, this classification is constantly changing, and is subject to many variables such as the size and position of the domesticated within the field of domestication. One rabbit in one's backyard can be classified as "cute," which can be reduced, as far as the logic of domestication is concerned, to "very harmless" and aesthetically/emotionally pleasing (that is, useful) and can bring out our most hospitable self. Ten rabbits, however, would see the classification "cute" turn into "pest" and the exterminator in us quickly replaces our inclusionary impulses.

A similar logic can be seen to operate with what I have called "numerical racism." A typical outburst would go like this: "I had no problem with the Vietnamese family next door. When they bought their house and settled in we became good friends. The wife introduced us to many Vietnamese dishes and she even used to cook for us. But then, two brothers bought two other houses up the road. And then the uncle did. The street is turning into a Vietnamese street. You notice this even from the smells of the cooking when you are walking up the road. There's just too many of them now."

Like with the evolution from one rabbit to ten, the evolution from one Vietnamese to ten generates the same change in classification from the "harmless and useful" toward the "harmful and useless" end of the axis. There is however a different logic that is not numerical. This is when something is perceived in its essence to be harmful and useless, the way a snake is. Here, one is enough to bring out the exterminator in us. The Jew of the Nazis was like this. One would never hear a Nazi saying "one Jew was all right. Now there are too many." One Jew is always already too many. One of the interesting developments that one witnesses looking at Islamophobic classifications of Muslims from that perspective is that we precisely see a transformation from a numerical to an essentialist logic. While there are still many people who talk about Muslims in terms of numbers, there is nonetheless a marked shift and an increase in the people for whom one Muslim is already more than enough (see, for example, the *New York Times* article, "Latvia Wants to Ban Face Veils, for All 3 Women Who Wear Them" [Martyn-Hemphill 2016]).

To be sure, let us recall here that we are analyzing matters entirely from the perspective of the domesticating power. The classifications "useful" and "harmful" are clearly from the perspective of the domesticator. Mary Midgley (1978: 359) has taken Iris Murdoch to task for writing about "the pointless existence of animals." To Midgley, "calling the bird's existence pointless means only that it is not a device for any human end." When something or someone metaphorically is classified as a "weed" (that is, useless and harmful) what it means is "something we have no use for that is competing with us for nutrition in the soil." A similar principle is behind the notion of "pest" and other such terms. Exterminatory tendencies are always directed toward what we have called in a previous chapter the "ungovernable," which we might now call the "undomesticable." The undomesticable like the weed, or the wolfish Muslim of the Islamophobe, is an entity whose "being" has successfully resisted numerous attempts at transforming it into a "being for" the domesticator. Such a figure of the undomesticable makes a fictional appearance in Rawi Hage's *Cockroach* (2009). It is a novel about a particularly unattractive Arab man. People look down on him with disgust as something abject. He turns into the cockroach that people metaphorically assume him to be. But instead of being inferiorized by the classification he uses it to empower himself, without becoming any more endearing. It's as if he is saying, I am repulsive, there is nothing about me to "tolerate," there is nothing valuable about me that you can be multicultural about, and yet, I am here to stay.

From elementary to complex domestication

While the above gives us the broad logic behind the generation of domesticating practices, it is important to finish by highlighting that such a broad logic, revealing as it may be, also conceals some important complexities. A good entry point into these is to note that domestication hardly ever starts from scratch. We are all born into always already domesticated spaces. As such we are all born into spaces that are already occupied, into symbolic spaces where the processes of polarization are already operating and have had a crucial impact on social reality, and into spaces where the forces aiming at positioning various forms of otherness are already in place, even if they are also continually challenged or in crisis.

What are the ramifications of this preexistence of domesticated spaces? The first is that domesticated spaces are not created by some kind of master domesticator ruling above everyone else. A domesticated space is a space where, to use Freud's imagery, the father of the horde has already been assassinated, laying the ground for the institution of his domesticating law. Domesticated spaces institute the rule of the law of the father rather than the rule of the father. This means that rather than a single domesticator we have a multiplicity of subjects legitimized to speak and domesticate in the name of the law of the father. An important dimension of such a world is that it is not neatly divided between domesticators and domesticated. Even within a patriarchal household, we cannot say that the patriarch is the domesticator and everyone else is the domesticated: women have domesticating power over children, children have domesticating power over servants and pets, and so on. A more complex understanding requires us to recognize the way social systems involve an unequal distribution of domesticating power within the same domesticating field. We can even say that a domesticated space involves an unequal distribution of subjectness and objectness. Everyone is to a certain extent part-domesticated, part-domesticator, but not everyone is so in the same way or to the same degree.

In "Techniques of the Body," Marcel Mauss gives us one of the earliest theorizations of this process:

> The techniques of the body can be classified according to their efficiency, i.e. according to the results of training. Training, like the assembly of a machine, is the search for the acquisition of an efficiency. Here it is a human efficiency. . . . These procedures that we apply to animals men voluntarily apply to themselves and to their children. The latter are probably the first beings to have been trained in this way, before all the animals, which first had to be tamed. (Mauss 1973: 77–8)

This absence of a neat divide between domesticators and domesticated does not signify that the relation domesticator-domesticated is no longer operative as a structuring principle.

Finally, to say that one is born in an already domesticated space also means that one does not create the classifications of the domesticated through a continual act of "first contact." Degrees of usefulness and harmfulness associated with certain positions and techniques of positioning are mostly inherited. The emphasis then has to turn toward such modes of inheritance. The ecological crisis just as much as the racial/colonial crisis we have been examining can well be understood as a crisis of inheritance, where the categories and the positions that are colonially inherited are out of sync with the way reality is evolving.

Negotiating the wolf

When examining the polarization of difference in the previous chapter, I highlighted the fact that it is part and parcel of generalized domestication as a mode of existence. It is both the product and the condition of possibility of domesticating practices. That this polarization is an integral part of a practice is important to highlight in the face of a long tradition of criticism that claims to want to "subvert" and "undermine" dualism, as if dualism was merely a form of thinking that took over the world because the publishers of Descartes's *Meditations* ensured that every household has a copy that is read to infants from the moment of their birth, imbuing them with faulty dualistic thinking. Dualism and polarization are part and parcel of generalized domestication as a mode of existing in the world and it is highly unlikely that a mode of existence so closely articulated to the realm of human necessity will cease to exist: for as long as beings, whether humans, animals, or plants, need to ensure their survival in the world, there will be domestication.

Even Val Plumwood, who has offered one of the more sophisticated analyses of "dualism," ends up seeing it as "a way of thinking about the relation between humans and nature" rather than as an intrinsic component of a practice. She begins by providing an account of dualism that is close to what we have called polarization. She argues that "dualism is the process by which contrasting concepts (for example, masculine and feminine gender identities) are formed by domination and subordination and constructed as oppositional and exclusive" (Plumwood 1993: 31). Dualism, she argues,

> is a process in which power forms identity, one which distorts both sides of what it splits apart, the master *and* the slave, the coloniser *and* the colonised, the sadist *and* the masochist, the egoist *and* the self-abnegating altruist, the masculine *and* the feminine, human *and* nature. (p. 32)

She even makes a point of highlighting that this dualist logic is "formed from a necessity inherent in the dynamic and logic of domination between self and other, reason and nature" (p. 4). This, in fact, is her main thesis. Yet it appears as if Plumwood perceives this logic as independent somehow of the practices of domination she analyzes so well. Her critique ends up with dualism perceived as a logical defect that one can fight on the terrain of philosophy and logic regardless of the practical necessity that she herself recognizes. Thus, she argues that the problem lies in the way "western culture has treated the human/nature relation as a dualism" (p. 2). The solution becomes an intellectualist one. Indeed she even proposes "some remedies for overcoming dualised identity" (p. 42). These involve the "redefinition or reconstruction in less oppositional and hierarchichal ways" of reason, science, and individuality (p. 4).

More recently, Aaron Gross, introducing his edited volume, *Animals and the Human Imagination: A Companion to Animal Studies*, continues the same tradition of argumentation against polarization, dualism and binaries. He argues that "highlighting the 'co-construction' of the categories human and animal is an attempt to question dominant Western articulations of the human/animal binary that overwhelmingly view this division of the world into human and animal as a fact. Far from being a datum given in the natural order, the human/animal binary has always been and remains unstable, disputed and negotiated" (Gross 2012: 2). Against this tendency to oppose social construction and facts, my argument in the previous chapter aimed to emphasize that seeing in humans and animals polarized constructions doesn't make them less of a fact. It is simply incorrect to equate a fact with what is classified as "natural." It is like saying that a chair is less of a "fact" than a tree because a chair is a social construction while the tree is "natural."

There is a good reason for us to dwell on this seemingly abstract issue at this concluding stage of the book. Important practical and political questions lie behind differing interpretations of "dualism" or "polarization." Everyone in this debate agrees that there are logics of dualism behind the ecological and colonial crises. Everyone also agrees that there is a need for a critique of dualism as part of seeing our way out of these crises. But not everyone agrees about what such a critique entails. Among those who see their task to be a "subversion" of dualistic thinking, dualism is a bad mode of thinking. We ought to get rid of it and struggle to think our relation to the human or animal other differently. I, on the other hand, want to emphasize dualism as integral to a practical mode of being. This clearly includes thinking but insisting on seeing it as one among many other components of practice. Somehow, I don't think many would disagree with this, not even those who highlight dualism as a mode of thinking. It is more the question of following up and being consistent about what this means. And one of the most important things that it means is that one challenges dualism by challenging generalized domestication as a practice, not just by picking on the "thinking" part of it.

But there is another point that we have already raised and that makes such an argument difficult. It is the idea that dualism is an integral, and even universal, dimension of how we exist in the world. It is not bad as such, nor is it likely to disappear in any case. This makes the idea of subverting domestication and dualism practically nonsensical. Understandably, some might think that I am painting myself into a corner: if one argues as I have that generalized domestication with its dualistic luggage has been generative of the colonial/racial and ecological crises, and then proceeds to argue that generalized domestication is inevitable, isn't one already admitting the impossibility of change? In this conclusion, I want to argue that this is not the case. Instead I propose that the problem is not generalized domestication as such. Rather, it is the way modernity has favored generalized domestication to the extent that, for a long time now, it has let it monopolize the scene. It is perceived as if there are no other modes of inhabitance and enmeshment beside it. Our aim, therefore, should not be "antidomestication" or "antidualism"; rather, it should be to oppose their dominance and monopoly. We thus need to aim for a recovery of the multiplicities of modes of inhabitance that capitalist modernity has excluded and marginalized. Anthropological research has a particularly important role to play here.

Outside generalized domestication

Since its inception, and notwithstanding its now well-criticized entanglement with colonialism, one of anthropology's core critical impulses has been, as Stanley Diamond has put it, to "giv(e) us a glimpse of another human possibility" (1974: xvi), bringing forth what Elizabeth Povinelli (2012) calls "the otherwise." Critical anthropologists researching radically different cultural spaces did not merely aim to discover different modes of being, inhabitance, and enmeshment. They aimed to do so with a critical eye on their own culture, where invariably generalized domestication dominated. The otherness discovered "over there" invited a challenge to see that same otherness lurking in the cracks of the dominant mode of existence "over here." This meant that for such critical anthropologists, no matter how dominant generalized domestication, capitalism, and modernity are in the West, there was always an outside to this dominance: modes of being that exist in a minor, marginal, or repressed (but in any case less visible) way beneath or around or in the cracks of the dominant mode of life. In this sense, every anthropology of elsewhere was an archaeology of one's own society, unearthing within it those more or less hidden traces of other modes of living and relating.

The two most important modes of being that anthropology has helped elucidate are what we can call the "reciprocal" and "mutualist" modes of existence. The first is grounded in the logic of gift exchange as analyzed initially and classically in the work of Marcel Mauss on the subject (1990 [1925]). The second emerges from the study of animism, beginning with the classic work of E. B. Tylor, *Primitive Culture* (1976 [1871]), and Lucien Lévy-Bruhl (1985) on the concept of participation in "mystical mentalities," and more recently, Marshall Sahlins's analysis of the logic of "mutuality" in kinship relations (2013). I will briefly explain what each of these modes of being entails, the way they differ from generalized domestication, and why they are important to us.

The mutualist mode of existence is also about interexistence. I am borrowing the concept of mutuality here from recent work by Marshall Sahlins on kinship. It highlights an order of existence where people (and animals, plants, objects, and so on) exist in each other. As Lucien Lévy-Bruhl analyzed it, it is a mode of living and thinking where we sense ourselves and others as "participating" in each other's existence, where the life-force of the humans and the nonhumans that surround us is felt to be contributing to our own life-force. If the domesticating mode of existence stresses a sense of boundaries concerned with the delineation of a space of sovereignty, the mutualist mode of existence underscores a reality where boundaries between self and other, human and animal, and so on, are far less absolute and even nonexistent, and where we experience an interpenetration between self and other. The idea that the being of others "participates" in my being is only present in generalized domestication in the process of extraction. In mutualism it is given in the very nature of being. A very mundane experience can communicate this: I walk my dogs in the park. As they run around and bounce happily I feel my own being bouncing. The state of their being participates in the state of my being. The anthropologist Nurit Bird-David (1999), who has written an important work on animism, calls it "relational." But this gives the impression that only nice relations are relations, for it is not clear why exploitation and extraction are not equally relational. Furthermore, mutualism is perhaps the least relational mode of existence insofar as relations assume two separate entities exterior to each other relating, and as Lévy-Bruhl emphasized, the entities that make up the mutualist mode of existence are the least separately delineated.

Reciprocal modes of existence on the other hand do have a clear sense of a boundary between entities. However, unlike with generalized domestication, the boundary between them is not problematized primarily as one of sovereignty but as a point of contact and exchange. To put it schematically, if generalized domestication initiates a mode of being where otherness is always an otherness that is instrumentalized and perceived to exist "for me," and if the mutualist mode of existence points to a way of being in which otherness exists "in me," the reciprocal mode of existence highlights a dimension in which otherness exists "with me." This "withness" is the withness of the offering. The other is always already in a state of giftedness in relation to me.

The work of Mauss on the gift has generated more work than perhaps any other in anthropology. This work has been dominated by the differentiation between societies dominated by gift exchange and those dominated by capitalist commodification. More recently Anna Tsing (2013) has shown how capitalist commodification can actually harness gift economies. So there is a massive literature that aims to show how gift economies themselves are an extension of the very instrumental reason and calculative relationality that rules in the realm of domestication. Notwithstanding the importance of this "demystificatory" approach (perhaps Bourdieu's [1990] work is exemplary in this regard), the brilliance of Mauss's work was to insist that if one sees only this logic of self-interest one misses a different dimension, which is precisely that of the space of radical difference where *The Gift* takes us. In that space, things are not merely "offered" as gifts in a strategically motivated act. Rather, the reciprocal gift relation is a relation that is always a surplus to the instrumental calculative relation. In a sense, there is a condition of inter-giftedness between everything that exists which precedes gift exchange and that the gift affirms and brings forth (this is what goes missing in philosophical critiques of Mauss that want the gift to be totally free of an interest in relationality; see, for example, Derrida [1992]). People, animals, plants, and objects stand as gifts toward each other. It is one of their modes of existence. A good way to exemplify this is to see people's reaction to a child entering a room: people open up to the child's presence as an offering; its mere appearance is treated as a gift. Some forms of religious thought equally capture this dimension by making life and everything that exists a gift of God requiring to be recognized as such, and where the very recognition of this giftness, the sentiment of gratitude, is itself a form of reciprocating the gift. Mauss's analysis of gift exchange showed that while this reciprocal gift-based mode of existence is minor in our society and indeed tends to become negligible with the increased dominance of calculative instrumental logic, there are societies where this dimension remains far more pronounced.

As indicated above, what is important to stress is that for anthropologists such as Mauss and Lévy-Bruhl and many others, it is never the case that "participatory" or "gift" logic is happening "over there" while generalized domestication and instrumental logic are here. Mauss makes it clear from the start that he is concerned with a search for the existence of gift exchange in his own society. Likewise Lévy-Bruhl couldn't be clearer in arguing that the peoples among whom the mystical/participatory mentality dominates are never just mystical.

> In practice, they act as if they have full confidence, just as we do, in the constancy of the laws of nature and the permanence of the forms of living beings: they need to in order to survive and their techniques are witness to this. (Lévy-Bruhl, 1985: 130)

The same goes for our societies: "there is a mystical mentality that is more easily demarcated and therefore more easily observable among the 'primitives' than in our societies, but it is present in every human mind" (Lévy-Bruhl, 1985: 131). Thus for Lévy-Bruhl social life always contains what he calls an "*enchevêtrement de réalités*," an entanglement of realities: no social space is exclusively a relation of generalized domestication, or a mutualist or a reciprocal relation. I might face the tree as a domesticator and evaluate how useful it is for me, I might even cut it down because I need fire or to build a house. Still, this does not mean that this tree did not present itself to me as a gift or that I experienced it as a gift at the same time, or that I even looked at it and felt more alive at the mere fact that it existed next to me. All social relations are an entanglement of multiplicities. It is the recovery of this multiplicity to temper the overdomination of generalized domestication that we need today to oppose the proliferation of racist and ecologically destructive modes of domination.

Ethnographies of the northern hunters are a good example of such a coexisting multiplicity. As Nadasdy explains:

Anthropologists have long been aware that many northern hunting peoples conceive of animals as other-than-human persons who give themselves to hunters. By accepting such gifts from their animal benefactors,

> hunters incur a debt that must be repaid through the performance of certain ritual practices. (Nadasdy 2007: 25)

Hunting in itself presumes generalized domestication as a mode of existence since it will necessarily involve a mode of capture that necessitates an instrumental evaluation of "what is good to hunt and eat," fat versus lean, tough versus soft, and so on. But as Nadasdy emphasizes, for the northern hunters: "Hunting in such societies should not be viewed as a violent process whereby hunters take the lives of animals by force. Rather, hunting is more appropriately viewed as a long-term relationship of reciprocal exchange between animals and the humans who hunt them" (p. 25). To see the animal as gift involves a different mode of othering, while negotiating the need for hunting with the giftedness of the animal has important ecological effects on the intensity of hunting.

Generalized domestication: from symbolic violence to orthodoxy

Examining both the possibilities of other modes of existence and, as important, the possibility of negotiating an entanglement of a multiplicity of modes of existence, serves to highlight the poverty of our mono-relational present. Generalized domestication, instrumental logic, and dualism are hardly the creation of Western modernity. Nor is "exploitation" invented by capitalism. I am in agreement with Nietzsche (2002: 153) when he says that the idea of removing exploitation is like a promise "to invent a way of life that would dispense with all organic functions." Exploitation, as he argued, does not belong to a corrupt system, "it belongs to the essence of being alive." Where I have parted company with him is when he concludes that exploitation "is a result of the genuine will to power, which is just the will of life." What we are trying to argue instead is that there are many other ways of "willing life" that are not the expression of an exploitative will to power. And it is toward this multiplicity that we need to orient ourselves to temper the exploitative will to power.

Nietzsche naturalizes Western modernity's eclipsing and marginalization of other "wills to life," such as the modes of being we referred to as mutualism and reciprocity. During modernity's reign, such modes were tolerated and legitimized as pertaining to the worlds of artists, poets, and other fantasists but not to the serious world of political and economic life. This is what we need to aspire to today. Our impasse is not the product of the dualist mode of thinking, nor is it only the problem of capitalist exploitation or the domesticating mode of existence. Rather, it is the problem of the mono-realism that capitalist modernity has locked us into: the fact that we are not able to think of solutions, or worse, we are unable to ask questions, other than through and within the categories of generalized domestication. The recurring questions that our societies keep asking bear the mark of this mono-realist impasse well before they are answered: How are we to manage nature? How are we to manage the ungovernable Muslim? It is within this restricted frame that we keep generating destructive solutions driven by a domesticating desire to "recover" an omnipotence we never had, whether they are of the order of geo-engineering or of the order of the fantasies of extermination that are generated in the encounter with the Muslim wolf. The question, "Is it possible not to consider nature or the Muslim as a managerial problem?" does not come to mind in the dominant governmental milieus. And yet this is perhaps one of the most urgent questions we are facing: is it possible to forefront other modes of inhabitance and relationality so as to relate with nature and the Muslim other in a way that is not exclusively managerial? Are exterminatory fantasies a necessary by-product of the way we relate to our metonymic and metaphoric wolves?

To a certain extent it can be said that the very fact that many today are asking these questions demonstrates the erosion of the hold that generalized domestication has had on us all. In Pierre Bourdieu's theorization of cultural domination he usefully differentiates between "symbolic violence" and "orthodoxy." A state of symbolic violence is a state where a cultural form dominates others opposing it so completely that the struggle between them is forgotten. The opposition becomes so minimal that people consider the existing state of affairs as "natural," as "something that goes without saying." A state of "orthodoxy" is still a state where a cultural form overwhelmingly dominates but its domination is visible, since its very existence as an orthodoxy signifies the existence of an equally visible or pronounced heterodoxy.

It can be said that today generalized domestication has moved, under the effect of the ecological crisis, from being a mode of symbolic violence to being an orthodoxy. Everywhere, counterhegemonic voices and counterhegemonic practices are emerging and growing stronger. Technologies that work through negotiation rather than solely through extraction are becoming increasing important, from the resurrected bicycle to wind farms to more negotiated modes of milking cows. In farming, cooking, and eating as well as modes of hiking (such as minimum-impact bushwalking, which invites a heightened consciousness of what one is disturbing in the process of walking in the bush) a logic of negotiated being infused with reciprocity and mutualism is emerging everywhere, creating wider and wider networks of alternative modes of existence.

Perhaps one of the most remarkable voices of opposition to have emerged in recent times has been Pope Francis and his *Encyclical on the Climate Change and Inequality* (2015). In it the pope begins precisely by recognizing the end of domestication's symbolic violence. As he put it:

> Following a period of irrational confidence in progress and human abilities, some sectors of society are now adopting a more critical approach. We see increasing sensitivity to the environment and the need to protect nature, along with a growing concern, both genuine and distressing, for what is happening to our planet. (Pope Francis 2015: 14)

Like an anthropologist highlighting the possibility of a multiplicity of modes of enmeshment, he argues that the problem is the mono-realism associated with generalized domestication:

Men and women have constantly intervened in nature, but for a long time this meant being in tune with and respecting the possibilities offered by the things themselves. It was a matter of receiving what nature itself allowed, as if from its own hand. Now, by contrast, we are the ones to lay our hands on things, attempting to extract everything possible from them while frequently ignoring or forgetting the reality in front of us. (p. 67)

Furthermore, as if writing specifically to help me finish this book on a good religious note, the pope emphasizes the interconnection between the ecological and the social, pointing out that "the human environment and the natural environment deteriorate together; we cannot adequately combat environmental degradation unless we attend to causes related to human and social degradation" (p. 29).

The pope also emphasizes the order of the gift, which is not surprising given the emphasis on the giftedness of the earth in Christian theology: "The destruction of the human environment is extremely serious, not only because God has entrusted the world to us men and women, but because human life is itself a gift which must be defended from various forms of debasement"

(p. 5). And for good measure, he also mobilizes St. Francis, who in the language we have developed earlier is a quintessential "mutualist":

> Saint Francis helps us to see that an integral ecology calls for openness to categories which transcend the language of mathematics and biology.... His response to the world around him was so much more than intellectual appreciation or economic calculus, for to him each and every creature was a sister united to him by bonds of affection. (pp. 19–20)

From the ethical to the political

We can see from the above that, at the very least, there are practices and voices growing in number and importance involved in a critical engagement with the dominance of generalized domestication. These voices are increasingly offering at least a marginal moral challenge to what is really a straitjacketing by generalized domestication of the imagination necessary to handle the crises produced by this very straitjacketing. But is it the case that "morality" and "ethical discourse" must always be marginal? Teresia Teaiwa describes listening to Kiribati's President Anote Tong at a conference on climate change at Victoria University of Wellington, and being surprised to hear him equating climate change with the Trans-Atlantic slave trade:

> Slavery, he said, was a system that was justified solely by its profitability. Morality was all that opponents of slavery had to argue against slave plantation economics.
>
> This is the same with climate change, Tong argued. Climate change is the consequence of a system justified solely by its profitability. The fossil fuel and coal industries, for example, are profitable. But they're also immoral. They're reaping profits for the few, while spreading the costs around the world. And some of these costs include the loss of whole homelands and livelihoods. (Teaiwa 2016)

This is also true of colonial racism as I presented it here. The colonial instrumental economistic logic that justifies Australia's inhumane detention centers, for example, cannot be opposed on its own pragmatic grounds. The detention centers fed by Islamophobia do their job of stopping the arrival of asylum seekers by boat very well. The opposition to them in Australia as elsewhere is largely moral. Critics often say, "It is working but what is it doing to us?" This raises the question of the power of "the moral and the ethical" in opposing "the profitable" and the "instrumental." To what extent can morality be transformed into a potent political force? As we have argued, one of the key characteristics of capitalism, but also more generally of generalized domestication, is the continual oscillation between aggressive profiteering and domination and the production of "civilized homeliness." The pope's ethical discourse can be seen as a moment of homeliness in this oscillation, providing the usual corrective to the increased social, political, and economic nastiness and aggression we find ourselves in today. So is the function of the religious to be, as in Marx's old dictum, "the soul of a soulless world" which leaves the world fundamentally soulless? or is it a soul that actually challenges the soullessness? That is, is the ethical today the new political space for those seeking change, given how corrupt other political spaces have become?

In this book, by showing the way racism reproduces the attitudes and dispositions that are behind the ecological crisis, I have argued that no political force aiming for social change can ignore the unity of the principle of generation of both the racial and the ecological crisis. I have also argued that it is possible to formulate an alternative ethico-political direction to the dominance of generalized domestication that is not based merely on a "good idea" but on an already existing "practical ground" made out of the multiplicity of surviving forms of inhabitance and relationality. Whatever this direction might be, however, it cannot ignore the fundamental unity of the struggle for ecological change and against colonial racism.

Agamben, Giorgio. 1998. *Homo Sacer: Sovereign Power and Bare Life.* Stanford University Press.

Agamben, Giorgio. 2005. *State of Exception.* The University of Chicago Press.

American Interest. 2015. "Europe under Siege." August 3. Accessed October 29, 2015. http://www.the-american-interest.com/2015/08/03/europe-under-siege/

Arata, Stephen. 1996. *Fictions of Loss in the Victorian Fin de Siècle: Identity and Empire.* Cambridge University Press.

Arendt, Hannah. 1944. "Race-Thinking Before Racism," *The Review of Politics,* 6, 1: 36–73.

Arjana, Sophia Rose. 2015. *Muslims in the Western Imagination.* Oxford University Press.

Aristotle. 2014. *Complete Works of Aristotle, Volume 2: The Revised Oxford Translation,* ed. Jonathan Barnes. Princeton University Press.

Armstrong, Philip. 2008. *What Animals Mean in the Fiction of Modernity.* Routledge.

Arrighi, Giovanni. 2009. "The Winding Paths of Capital: Interview by David Harvey," *New Left Review,* 56 (March-April): 61–94.

Backhouse, Maria 2015. "Green Grabbing – the Case of Palm Oil Expansion in So-Called Degraded Areas in the Eastern Brazilian Amazon," in *The Political Ecology of Agrofuels,* ed. Kristina Dietz, Bettina Engels, Oliver Pye, and Achim Brunnengräber, pp. 167–85. Routledge.

References

Benveniste, Emile. 1973. *Indo-European Language and Society*. University of Miami Press.

Bird-David, Nurit. 1999. " 'Animism' Revisited: Personhood, Environment, and Relational Epistemology," *Current Anthropology*, 40, 1, pp. 67–91.

Bökönyi, Sandor. 1989. "Definitions of Animal Domestication," in *The Walking Larder: Patterns of Domestication, Pastoralism and Predation*, ed. Juliet Clutton-Brock, pp. 22–27. Routledge.

Bourdieu, Pierre. 1990. *The Logic of Practice*. Polity Press.

Bourdieu, Pierre. 2000. *Pascalian Meditations*. Polity Press.

Bradiotti, Rosi. 2013. *The Posthuman*. Polity Press.

Brague, Rémi. 2007. *The Law of God: The Philosophical History of an Idea*. The University of Chicago Press.

Brown, Wendy. 2010. *Walled States, Waning Sovereignty*. Zone Books.

Bullard, Robert D., ed. 1993. *Confronting Environmental Racism: Voices from the Grassroots*. South End Press.

Carbone, Geneviève. 1991. *La Peur du loup*. Gallimard.

Cárdenas, Roosbelinda. 2012. "Green Multiculturalism: Articulations of Ethnic and Environmental Politics in a Colombian 'Black Community,' " *Journal of Peasant Studies*, 39, 2: 309–33.

Clark, Brett, and Foster, John Bellamy. 2009. "Ecological Imperialism and the Global Metabolic Rift: Unequal Exchange and the Guano/Nitrates Trade," *International Journal of Comparative Sociology*, 50, 3–4: 311–34.

Corcoran, Ann. 2015. "More Evidence Europe Is under Siege by the Flow of Illegal Migrants from Africa." Refugee Resettlement Watch website, May 31. https://refugeeresettlementwatch.wordpress.com/2015./05/31/more-evidence-europe-is-under-siege-by-the-flow-of-illegal-migrants-from-africa/.

Deleuze, Gilles. 1988. *Spinoza: Practical Philosophy*. City Lights Books.

Descola, Philippe. 2014. *Beyond Nature and Culture*. The University of Chicago Press.

Derrida, Jacques. 1992. *Given Time. I, Counterfeit Money.* The University of Chicago Press.

Derrida, Jacques 2008. *The Animal that Therefore I Am.* Fordham University Press.

Diamond, Stanley. 1974. *In Search of the Primitive: A Critique of Civilization.* Transaction Publishers.

Digard, Jean-Pierre. 1990. *L'Homme et les animaux domestiques: Anthropologie d'une passion.* Fayard.

Douglas, Mary. 1966. *Purity and Danger: An Analysis of Concepts of Pollution and Taboo.* Routledge & Kegan Paul.

Ducos, Pierre. 1989. "Defining Domestication: A Clarification," in *The Walking Larder: Patterns of Domestication, Pastoralism and Predation*, ed. Juliet Clutton-Brock, pp. 28–30. Routledge.

Easterman, Daniel 1987. *The Seventh Sanctuary.* Doubleday.

Eckersley, Robyn. 2015. "Democracy in the Anthropocene." Paper presented at the symposium "Quelles transitions écologiques?" Centre culturel international de Cerisy-la-Salle, France, July 1.

Elias, Norbert. 1994. *The Civilizing Process: The History of Manners, and State Formation and Civilization.* Blackwell.

Esposito, John, and Kalin, Ibrahim. 2011. *Islamophobia: The Challenge of Pluralism in the 21ˢᵗ Century.* Oxford University Press.

Foucault, Michel. 1979. *Discipline and Punish: The Birth of the Prison.* Vintage Books.

Foucault, Michel. 1991. "Governmentality," in *The Foucault Effect: Studies in Governmentality: With Two Lectures by and an Interview with Michel Foucault*, ed. Graham Burchell, Colin Gordon, and Peter Miller, pp. 87–104. The University of Chicago Press.

Foucault, Michel. 2008. *The Birth of Biopolitics: Lectures at the Collège de France, 1978–1979.* Palgrave Macmillan.

Gatens, Moira. 1996. *Imaginary Bodies: Ethics, Power and Corporeality.* Routledge.

God's Kingdom Ministries. 2015. "Reverse Colonialism and Its Cause." September 17. Accessed October 5, 2015. http://gods

References

-kingdom-ministries.net/daily-weblogs/2015/09-2015/reverse
-colonialism-and-its-cause/

Goldberg, David Theo. 2015. *Are We All Post-Racial Yet?* Polity.

Gros, Daniel. 2015. "Europe under Siege?" Project Syndicate website, September 8. Accessed October 29, 2015. https://www.project-syndicate.org/commentary/european-asylum-burden-sharing-by-daniel-gros-2015-09

Gross, Aaron. 2012. "Introduction and Overview: Animal Others and Animal Studies," in *Animals and the Human Imagination: A Companion to Animal Studies*, ed. Aaron Gross and Anne Vallely, pp. 1–23. Columbia University Press.

Guillaumin, Colette. 1995. *Racism, Sexism, Power, and Ideology.* Routledge.

Hage, Ghassan. 1996. "The Spatial Imaginary of Nation-Building: Dwelling-Domesticating/Being-Exterminating," *Environment and Planning D: Society and Space*, 14, 4: 463–485.

Hage, Ghassan. 2000. *White Nation: Fantasies of White Supremacy in a Multicultural Society.* Routledge.

Hage, Ghassan. 2003. *Against Paranoid Nationalism: Searching for Hope in a Shrinking Society.* Pluto Press.

Hage, Rawi. 2009. *Cockroach.* W.W. Norton & Co.

Hannerz, Ulf. 1996. *Transnational Connections: Culture, People, Places.* Routledge.

Haraway, Donna. 2008. *When Species Meet.* University of Minnesota Press.

Herzl, Theodor. *The Jewish State: An Attempt at a Modern Solution of the Jewish Question.* Pordes, 1972.

Houellebecq, Michel. 2015. *Submission: A Novel.* Farrar, Straus and Giroux.

Ken's Blog. 2009. "Reverse Colonialism." September 28. Accessed October 5, 2015. http://blog.kenperlin.com/?p=2162.

Korol, Tabitha. 2014. "A Pack of Wolves," *Dr. Rich Sweier*, December 22. http://drrichswier.com/2014/12/22/pack-known-wolves/

References

Kristeva, Julia. 1982. *Powers of Horror: An Essay on Abjection.* Columbia University Press.

Latour, Bruno. 2015. *Face à Gaïa.* La Découverte.

Le Bon, Gustave. 1894. *Les lois psychologiques de l'évolution des peuples.* F. Alcan.

Lévi-Strauss, Claude 1966. "Anthropology: Its Achievements and Future," *Current Anthropology*, 7, 2: 124–7.

Lévy-Bruhl, Lucien. 1985. *How Natives Think.* Princeton University Press.

Makdisi, Saree. 2014. *Making England Western: Occidentalism, Race, and Imperial Culture.* The University of Chicago Press.

Mallette, Karla. 2010. *European Modernity and the Arab Mediterranean: Toward a New Philology and a Counter-Orientalism.* University of Pennsylvania Press.

Marx, Karl. 1976. *Capital, Volume 1.* Penguin Books.

Martyn-Hemphill, Richard. 2016. "Latvia Wants to Ban Face Veils, for All 3 Women Who Wear Them," *New York Times*, April 19. http://www.nytimes.com/2016/04/20/world/europe/latvia-face-veils-muslims-immigration.html?_r=0

Mauss, Marcel. 1973. "Techniques of the Body," *Economy and Society*, 2, 1: 70–88.

Mauss, Marcel. 1990. *The Gift: The Form and Reason For Exchange in Archaic Societies.* Routledge.

Mbembe, Achille. 2003. "Necropolitics," *Public Culture*, 15, 1: 11–40.

McBrien, Justin. 2016. "Accumulating Extinction: Planetary Catastrophism in the Necrocene," in *Anthropocene or Capitalocene? Nature, History, and the Crisis of Capitalism*, ed. Jason W. Moore, pp. 116–37. PM Press.

Memmi, Albert. 1965. *The Colonizer and the Colonized.* Grossman.

Midgley, Mary. 1978. *Beast and Man: The Roots of Human Nature.* Cornell University Press.

Mitchell, Timothy. 2013. *Carbon Democracy: Political Power in the Age of Oil.* Verso.

References

Miles, Robert. 1993. *Racism After "Race Relations."* Routledge.

Moore, Jason W. 2011. "Ecology, Capital, and the Nature of Our Times," *Journal of World-Systems Analysis*, 17, 1: 108–47.

Nadasdy, Paul. 2007. "The Gift in the Animal: The Ontology of Hunting and Human-Animal Sociality," *American Ethnologist*, 34, 1: 25–43.

Nealon, Jeffrey. 2016. *Plant Theory: Biopower and Vegetable Life.* Stanford University Press.

Nietzsche, Friedrich. 1997. *On the Genealogy of Morality.* Cambridge University Press.

Nietzsche, Friedrich. 2002. *Beyond Good and Evil.* Cambridge University Press.

Norman, Daniel. 1980. *Islam and the West, the Making of an Image.* Edinburgh University Press

Novek, Joel. 2012. "Discipline and Distancing: Confined Pigs in the Factory Farm Gulag," in *Animals and the Human Imagination: A Companion to Animal Studies*, ed. Aaron Gross and Anne Vallely, pp. 121–51. Columbia University Press.

Palsson, Gisli. 1996. "Human-Environmental Relations: Orientalism, Paternalism and Communalism," in *Nature and Society: Anthropological Perspectives*, ed. Phillipe Descola and Gisli Palsson, pp. 63–81. Routledge.

Perbner, Pnina. 2005. "Islamophobia: Incitement to Religious Hatred – Legislating for a New Fear?," *Anthropology Today*, 21, 1: 5–9.

Plumwood, Val. 1993. *Feminism and the Mastery of Nature.* Routledge.

Policante, Amedeo. 2015. *The Pirate Myth: Genealogies of an Imperial Concept.* Routledge.

Pope Francis. 2015. *Encyclical on the Climate Change and Inequality: On Care for Our Common Home.* Melville House.

Povinelli, Elizabeth. 2012. "The Will to Be *Otherwise* / The Effort of Endurance," in *South Atlantic Quarterly*, 111, 3: 453–75.

Razack, Sherene. 2008. *Casting Out: The Eviction of Muslims from Western Law and Politics.* University of Toronto Press.

References

Reid, Sue. 2015. "Germany in a State of Siege." UK *Daily Mail*, September 26. http://www.dailymail.co.uk/news/article-3249667/Germany-state-SIEGE-Merkel-cheered-opened-floodgates-migrants-gangs-men-roaming-streets-young-German-women-told-cover-mood-s-changing.html.

Rodinson, Maxime. 1978. *La fascination de l'Islam*. La Découverte.

Rogers, Barbara. 1980. *The Domestication of Women: Discrimination in Developing Societies*. St. Martin's Press.

Saadi, Saïd Amirouche. 2010. *Une Vie, Deux Morts, Un Testament: Une Histoire Algérienne*. Harmattan.

Sahlins, Marshall. 2013. *What Kinship Is – And Is Not*. University of Chicago Press.

Said, Edward. 1978. *Orientalism*. Pantheon Books.

Shiva, Vandana. 1989. *Staying Alive: Women, Ecology, and Development*. Zed Books.

Shiva, Vandana. 1991. *The Violence of the Green Revolution: Third World Agriculture, Ecology, and Politics*. Zed Books.

Shryoc, Andrew. 2010. *Islamophobia/Islamophilia: Beyond the Politics of Enemy and Friend*. Indiana University Press

Spinoza, Baruch. 2005. *Ethics*. Penguin Books.

Sodahead. 2014. "Pack of wolves" [Internet comment by Temlakos~POTL~PWCM~JLA], December 22. http://www.sodahead.com/united-states/pack-of-wolves/question-4637542.

Spiegel, Marjorie. 1996. *The Dreaded Comparison: Human and Animal Slavery*. Mirror Books.

Steinmetz, George. 2007. *The Devil's Handwriting: Precoloniality and the German Colonial State in Qingdao, Samoa, and Southwest Africa*. University of Chicago Press.

Teaiwa, Teresia. 2016. "How Climate Change Is Like the Slave Trade," *E-Tangata: A Maori and Pasifika Sunday Magazine*, February 28. http://e-tangata.co.nz/news/how-climate-change-is-like-the-slave-trade

References

Temple, Julien. 2010. "Detroit: The Last Days." UK *Guardian*, March 11. http://www.theguardian.com/film/2010/mar/10/detroit-motor -city-urban-decline.

Thomas, Keith. 1983. *Man and the Natural World: Changing Attitudes in England 1500–1800*. Allen Lane.

Todorov, Tzvetan. 1994. *On Human Diversity: Nationalism, Racism, and Exoticism in French Thought*. Harvard University Press.

Tsing, Anna. 2013. "Sorting Out Commodities: How Capitalist Value Is Made Through Gifts," *Hau: Journal of Ethnographic Theory*, 3, 1: 21–43.

Tylor, Edward Burnett. 1976. *Primitive Culture: Researches into the Development of Mythology, Philosophy, Religion, Art, and Custom*. Gordon Press.

Uexküll, Jakob von. 2010. *A Foray into the Worlds of Animals and Humans*. University of Minnesota Press.

Viveiros de Castro, Eduardo. 2014. *Cannibal Metaphysics: For a Post-Structural Anthropology*. Univocal.

Washington, Booker T. 1963. *Up From Slavery: An Autobiography*. Doubleday.

Weiner, Annette B. 1992. *Inalienable Possessions: the Paradox of Keeping-While-Giving*. University of California Press.

Williams, Robyn. 2001. *2007 – A True Story, Waiting to Happen*. Hodder.

Wolfe, Patrick. 2006. "Settler Colonialism and the Elimination of the Native," *Journal of Genocide Research*, 8, 4: 387–409.

Zeuner, Frederick Everard. 1963. *A History of Domesticated Animals*. Hutchinson.

academic/scholastic perspective
  6–9
Agamben, Giorgio 52, 59, 67–8
American Interest 69–70
animal–human relationship
  domestication 85–6, 87, 88–90,
    98–9, 103
  hunting 123–4
  metaphors and similes 9–10,
    11, 32–3, 45, 46–8, 99–100
  polarization 114
  see also nature
anthropological perspective
  117–24
anti-Semitism 29–31, 32
apartheid 38, 39
Arata, Stephen 71, 73
Aristotle 17–18, 89
Armstrong, Philip 75, 76
assimilation 41–2
asylum seekers *see* migration

becoming-wolf of Muslim other
  29–37
Benveniste, Emile 91–2
biopolitics 52–3, 54, 55, 58, 62,
  65–6, 103–4
bodily techniques 99, 110
Bökönyi, Sandor 90

borders 37–41, 65
Bourdieu, Pierre 6, 101, 121,
  127

Cameron, David 35
capitalism *see* primitive
  accumulation
Carbone, Geneviève 35–6
class 19, 90
  racialized class borders 39–40,
    41
climate change 130–1
Cockroach (Hage) 108
colonialism
  borders 37–8, 40, 65
  classification 31–3
  and neo-colonialism 66–7
  plundering 57–60, 64–5
  polarization 100
  racism 27–9, 47–8, 50, 131–2
  reversal fantasies 68–77
Corcoran, Ann 68–9

Derrida, Jacques 25–6, 96
Descartes, René 112
detention centers, Australia 131–2
Diamond, Stanley 117
Digard, Jean-Pierre 87, 90, 96–7
dualism 112–17

Index

Easterman, Daniel 32–3
Elias, Norbert 66, 99
environmental crisis 1–3, 13–16,
    48–50, 62–3, 128–31
ethical and political discourses
    130–3
exploitation 53–4, 125
    see also generalized
        domestication; primitive
        accumulation
exterminability and
    ungovernability 77–81

Face à Gaïa (Latour) 76–7
Foucault, Michel 52, 79, 103–4

generalized domestication 85–7
    anthropological perspective
        117–24
    domestication of nature and
        88–94
    dualism and polarization 112,
        115–17
    elementary and complex
        108–11
    ethical and political discourses
        130–3
    as fantasy of viability 81–4
    process of occupation 94–7
    process of polarization 97–102
    spatial practices of extraction
        102–8
    symbolic violence and orthodoxy
        124–30
Genesis 84
gift exchange / giftedness 120–4,
    129–30
globalization 3–6

God's Kingdom Ministries 71–2
Goldberg, David 20
Gros, Daniel 70
Gross, Aaron 114
Guillaumin, Colette 20, 22–4

Hage, Rawi: *Cockroach* 108
Haraway, Donna 25–6
harm and usage classification
    104–6, 107, 110–11
Herzl, Theodor 95–6
Houellebecq, Michel: *Soumission*
    73–4
hunting practices 123–4

inferiority and inferiorization
    10–11
Islamophobia 3–6, 8–9, 12–13
    as colonial racism 27–9, 50
    generalized domestication
        106–8
    Muslim as wolf 33–7, 81, 96,
        126
    polarization 101–2
Israel 67–8, 95–6

Ken's Blog 72
Korol, Tabitha 35

Latour, Bruno: *Face à Gaïa* 76–7
laws, religious and national 43–5
Le Bon, Gustave 19
Lévi-Strauss, Claude 27–8
Lévy-Bruhl, Lucien 118–19,
    122–3

Makdisi, Saree 100
Marx, Karl 53, 54, 56–7, 132

Index

Mauss, Marcel 99, 110, 118, 120, 121, 122
Mbembe, Achille 52, 53
McBrien, Justin 63
Memmi, Albert 100
Middle East 37–8, 64, 65, 71–2
Midgley, Mary 107
migration (asylum seekers / refugees) 36, 40–1, 45, 48–9
    colonial reversal fantasies 68–77
Miles, Robert 18–19
multiculturalism 41–4
multiplicity 123–4
Muslim as wolf 33–7, 81, 96, 126
"mutual benefit" of domestication 89–91, 93
mutualist mode of existence 118–20, 130
mystical/participatory mentality 122–3

Nadasdy, Paul 123–4
national and religious laws 43–5
national borders 38–9, 40–1, 65
nature
    classification and domination 17–27
    domestication of 88–94
    domination and reversal fantasies 74–7
    see also animal–human relationship
Nealon, Jeffrey 52
Necrocene age 63
"necropolitics" 52–5, 58, 62, 64, 65–6
Nietzsche, Friedrich 12, 46–7, 82, 125
"numerical racism" 106–7

Obama, Barack 29
occupation, process of 94–7
orthodoxy and symbolic violence 124–30
other
    becoming-wolf of Muslim 29–37
    uncontainable 37–41
    unintegrable 41–4
overexploitation and overoscillation 61–8

patriarchy 11, 93–4, 109–10
Plumwood, Val 26, 113–14
polarization
    and dualism 112–17
    process of 97–102
Policante, Amedeo 59–60
political and ethical discourses 130–3
politics of ungovernable waste 46–51
Pope Francis 128–30, 132
Povinelli, Elizabeth 117
power relations 46–7, 90–1, 103–4
primitive accumulation 55–60, 85
    overexploitation and overoscillation 61–8

racialized class borders 39–40, 41
reciprocal mode of existence 118, 120–4
refugees see migration
Reid, Sue 69
religious and national laws 42–5
religious perspective 71–2, 84, 129–30, 132
Rushdie affair 44–5

Index

Sahlins, Marshall 118
scholastic/academic perspective
    6–9
sexage 20
Shiva, Vandana 21
slavery 28–9, 31, 90, 100, 131
social constructionism 97–8
Sodahead 35
Soumission (Houellebecq) 73–4
spatial practices of extraction
    102–8
speciesism 17–18
Spiegel, Marjorie 17
Spinoza, Baruch 63, 82
St. Francis 130
symbolic violence and orthodoxy
    124–30

Teaiwa, Teresia 130–1
Temple, Julien 74
terrorism 29–30, 33, 34–5, 45

Thomas, Keith 21, 86, 89,
    99–100
Tong, Anote 130–1
transnationalism 44–5

uncontainable other 37–41
ungovernability
    and exterminability 77–81
    politics of 46–51
unintegrable other 41–4
usage and harm classification
    104–6, 107, 110–11

Washington, Booker T. 100
Williams, Robyn: *2007* 75
wolf metaphor 33–7, 81, 96,
    126
Wolfe, Patrick 58, 61

Zeuner, Frederick Everard 86
Zionism 32–3, 67–8, 95–6